T0155197

Animal Welfare In Extensive Production Systems

The Animal Welfare Series

This is the first book in the *"Animal Welfare Series"* launched by *5M Publishing* in collaboration with the *Farm Animal Welfare Education Centre* (FAWEC) of the School of Veterinary Science at the Autonomous University of Barcelona (Spain).

The series covers current scientific knowledge on the welfare of all farmed species including both common species such as cattle, sheep, goats, pigs, poultry and fish and less common species such as camelids, deer, rabbits and horses. The series focuses on well referenced and peer reviewed academic material produced by established authorities.

The books in the series are authored and edited by a number of well recognized scientists in the field of animal welfare from a number of geographical origins and representing several different approaches and areas of expertise.

The level of the content reflects the diverse range of people interested in animal welfare. It is intended that the majority of the books in the series will be suitable for undergraduate level students of animal welfare within veterinary and animal science courses and on animal welfare specific courses at universities and agricultural colleges. Some books with a strong research emphasis will be aimed at the postgraduate level, whereas other books in the series will be more practically orientated and will be particularly useful for people who work daily with animals such as stockmen, herdsmen, transport workers and other animal handlers. Finally, we also hope that all books in the series will be of interest to practising veterinarians concerned with animal welfare, academic researchers, and people employed by NGOs, government agricultural employees and people affiliated with animal charities.

One of the main objectives of the series is to cover areas within the general field of farm animal welfare that until now have received less attention in the scientific literature. Animal welfare in extensive systems, which is the topic of this first book, is an example of this endeavour.

We very much hope that this new series will be a valuable contribution to the existing literature on farm animal welfare.

Xavier Manteca

Prologue

Rangelands represent the primary land type in the world. Animals that live in these vast and diverse ecosystems face multiple and different environmental challenges from those raised in intensive systems. Given this scenario, can we still apply a framework developed for animals raised in captivity to assess the welfare of free-ranging animals? How do animals cope with the imposed challenge of ever-changing and harsh conditions typically observed in rangelands? How can novel research and management approaches enhance the welfare of animals living in these environments? An attempt to answer these questions is what inspired the development of this multi-chapter book. In each chapter Authors explore key biotic and abiotic factors that challenge animals in extensive systems, describe adaptations to these challenges and propose future research and managerial directions aimed at improving the welfare of animals raised in these systems.

The Editors

Animal Welfare in Extensive Production Systems

The Animal Welfare Series

Editor: Juan J. Villalba

Series Editor: Xavier Manteca

First published 2016

Copyright © Juan J. Villalba 2016

Published by
5M Publishing Ltd,
Benchmark House,
8 Smithy Wood Drive,
Sheffield, S35 1QN, UK
Tel: +44 (0) 1234 81 81 80
www.5mpublishing.com

A Catalogue record for this book is available from the British Library

ISBN 978-1-910455-54-8

Book layout by Servis Filmsetting Ltd, Stockport, Cheshire
Printed by Replika Press Pvt Ltd, India
Photos as credited in the text

Contents

The Animal Welfare Series ii
Prologue iii
Contributors viii
Preface x

CHAPTER 1 Animal welfare in
extensive systems 1
  *Juan J. Villalba and Xavier
  Manteca*

CHAPTER 2 Thermal stress in
ruminants: Responses and
strategies for alleviation 11
  *Ahmed A.K. Salama, Gerardo Caja,
  Soufiane Hamzaoui, Xavier Such,
  Elena Albanell, Bauabid Badaoui,
  and Juan J. Loor*

CHAPTER 3 Adaptive responses
of rangeland livestock to manage
water balance 37
  Dean K. Revell

CHAPTER 4 Grazing and animal
distribution 53
  Derek W. Bailey

CHAPTER 5 Impacts of toxic
plants on the welfare of grazing
livestock 78
  *James A. Pfister, Benedict T.
  Green, Kevin D. Welch, Frederick
  D. Provenza, and Daniel Cook*

CHAPTER 6 Predation 103
  John Laundré

CHAPTER 7 The health of
livestock in extensive systems 132
  Pete Goddard

CHAPTER 8 Neonatal mortality
of farm livestock in extensive
management systems 157
  *Cathy M. Dwyer and Emma M.
  Baxter*

CHAPTER 9 Transport of livestock
from extensive production
systems 188
  Clive Phillips

Index 201

Contributors

Elena Albanell, Grup de Recerca en Remugants (G2R), Departament de Ciència Animal i dels Aliments, Universitat Autònoma de Barcelona, Spain

Bouabid Badaoui, Science Faculty, Mohamed V University, Rabat, Morocco

Derek W. Bailey, New Mexico State University, Las Cruces, USA

Emma M. Baxter, Animal and Veterinary Sciences, SRUC, Edinburgh, UK

Gerardo Caja, Grup de Recerca en Remugants (G2R), Departament de Ciència Animal i dels Aliments, Universitat Autònoma de Barcelona, Spain

Daniel Cook, USDA-ARS Poisonous Plant Research Laboratory, Logan, USA

Cathy M. Dwyer, Animal and Veterinary Sciences, SRUC, Edinburgh, UK

Pete Goddard, The James Hutton Institute, Aberdeen, UK

Benedict T. Green, USDA-ARS Poisonous Plant Research Laboratory, Logan, USA

Soufiane Hamzaoui, Grup de Recerca en Remugants (G2R), Departament de Ciència Animal i dels Aliments, Universitat Autònoma de Barcelona, Spain

John Laundré, State University of New York at Oswego, USA

Juan J. Loor, Department of Animal Sciences and Division of Nutritional Sciences, University of Illinois, Urbana-Champaign, USA

Xavier Manteca, School of Veterinary Science, Universitat Autònoma de Barcelona, Barcelona, Spain

James A. Pfister, USDA-ARS Poisonous Plant Research Laboratory, Logan, Utah, USA

Clive Phillips, Centre for Animal Welfare and Ethics, School of Veterinary Science, University of Queensland, Australia

Frederick D. Provenza, Department of Wildland Resources, Utah State University, Logan, Utah, USA.

Dean K. Revell, Revell Science, Duncraig, Western Australia and School of Animal Biology, The University of Western Australia, Nedlands, Western Australia

Ahmed A.K. Salama, Grup de Recerca en Remugants (G2R), Departament de Ciència Animal i dels Aliments, Universitat Autònoma de Barcelona, Spain and Sheep and Goat Research Department, Animal Production Research Institute, Giza, Egypt

Xavier Such, Grup de Recerca en Remugants (G2R), Departament de Ciència Animal i dels Aliments, Universitat Autònoma de Barcelona, Spain

Juan J. Villalba, Department of Wildland Resources, Utah State University, Utah, USA

Kevin D. Welch, USDA-ARS Poisonous Plant Research Laboratory, Logan, Utah, USA

Preface

From the standpoint of animal welfare, sustained attention has been given to intensive animal production systems, a pattern which contrasts with the little emphasis directed towards the welfare of animals living in extensive systems. However, the unpredictability and variability of environmental factors, paired with multiple managerial decisions dictated by humans constantly challenge animals living in natural landscapes, making the understanding of animal welfare in extensive systems a timely and significant endeavor. In turn, knowledge regarding the impact of grazing animals on their feed and water resources is key for enhancing the sustainability and stewardship of the rangelands they inhabit. Authors in this book analyze some of the key challenges animals face in extensive conditions and present the latest discoveries and future research directions on the direct and indirect consequences that those challenges impinge on animal welfare. It is our hope that this book stimulates new philosophies, ideas, research and management approaches aimed at enhancing the ability of animals to adapt to the dynamic environmental conditions imposed by extensive systems.

The Editors
Juan J. Villalba
Xavier Manteca

chapter one

Animal welfare in extensive systems

Juan J. Villalba[1] and Xavier Manteca[2]

INTRODUCTION

Rangelands are the primary land type in the world with approximately 50 per cent of the total land surface area (FAO, 2011). These key ecosystems comprise grasslands, shrublands, woodlands, and deserts with the common characteristic of limited moisture, climatic extremes, and variability in soils and topography. Due to the diversity encountered in rangeland systems, several names have been used around the world to more specifically describe this land type including prairies, plains, grasslands, steppes, pampas, scrublands, savannahs, mountain regions, deserts, semi-deserts, and arid lands (Lovina et al., 2009). Collectively, rangelands are important to society for the goods and services they produce and for the ecological services they provide (Society for Range Management). One of these services involves providing feed resources and water that nourish grazing and browsing animals. In turn, humans are dependent upon grazing animals for the provision of meat, milk, and fibre. Wild ruminants provide recreational opportunities like hunting and viewing, and domestic ruminants represent a fundamental source of power and transportation in many undeveloped countries (Van Soest, 1994). Given the global significance of animals living in extensive conditions, a clear understanding of the physiological and environmental factors underlying their welfare is essential. In turn, knowledge regarding the impact of ruminants on their feed and water resources as well as on soils is key for fostering the sustainable stewardship of rangeland landscapes.

It is clear that the environment poses multiple challenges to the welfare of animals and as stated by Broom (1986): "the welfare of an individual is its state as it regards its attempts to cope with its environment." Thus, welfare depends not only on whether the animal succeeds at coping with the challenges emerging from its environment, but also on whether coping attempts lead to

[1] Department of Wildland Resources, Utah State University, Utah, USA
[2] School of Veterinary Science, Universitat Autònoma de Barcelona, Barcelona, Spain

negative consequences for the animal. One of the most widely used conceptual frameworks to assess animal welfare in practice is that given by the United Kingdom Farm Animal Welfare Council, which states that the welfare of animals depends on whether they benefit from the so-called "five freedoms" (Farm Animal Welfare Council, 1992), namely: a) freedom from thirst, hunger, and malnutrition; b) freedom from discomfort; c) freedom from pain, injury, and disease; d) freedom to express normal behaviour; and e) freedom from fear and distress. It could be argued that rangelands should not compromise the five freedoms because animals are free to graze in their "natural" environment and as a consequence, should be able to cope with it. In fact, one approach to welfare considers that animals should be raised in a "natural" environment and allowed to behave in a "natural" way. People following this approach often study the behaviour of animals in the wild and compare it with that of animals raised in captivity (Duncan and Fraser, 1997). However, the environment is not static; on the contrary, it is highly variable and as a consequence, animals kept in the environment in which they have evolved can still face environmental challenges that exceed their capacity to cope (Villalba et al., 2016). This is – even when generally overlooked – a condition that constantly challenges the welfare of animals living in rangelands. Moreover, we submit that given the variability and unpredictability of changes occurring in rangelands, the challenges that compromise animal welfare under extensive conditions may in some instances be even greater than those observed for animals kept in captivity. Given this scenario, would it be possible to still apply the framework of the five freedoms designed for assessing the welfare of animals kept in captive environments to animals living in extensive conditions? In addition: How do animals cope with the imposed challenge of ever-changing conditions that are typically observed in rangelands? How can novel research and management approaches enhance the welfare of animals living in these environments?

An attempt to answer the aforementioned questions inspired the development of this book. Its structure relies on the identification of a key challenge (i.e., Chapter) faced by animals in rangelands followed by an exploration of the animals' "answers" to such challenge in terms of their behavioural and physiological adaptations. The key challenges and adaptations reviewed in this book involve: thermal stress (Chapter 2); water balance (Chapter 3); animal distribution (Chapter 4); toxic plants (Chapter 5); predation (Chapter 6); health (Chapter 7); neonatal mortality (Chapter 8); and transport (Chapter 9). Animals' responses to each of the challenges are presented and analyzed under the framework of the five freedoms. The authors then provide their own vision of the knowledge gaps which need to be filled with new research as well as potential future managerial directions aimed at improving the welfare of animals raised in rangelands. It is the hope of the editors that such exercise stimulates new ideas, research, and innovative practices aimed at providing animals with the means and conditions needed for an enhanced adaptation to the dynamic environmental forces that challenge animal welfare in extensive conditions.

CHANGE: THE ONLY CONSTANT

Temperature, Food, and Water

Rangelands are constantly changing. Some of these changes are inherent in the dynamic nature of the biotic and abiotic processes that underlie the functioning of rangelands. Some changes have only local effects but others have global effects and impacts like the recent increases in the levels of CO_2 in the atmosphere and their impact on whole ecosystems (IPCC, 2013). Climate variability can have significant negative effects on animals grazing in rangelands, particularly for dryland regions with low and variable precipitation and high temperatures in the growing season (Peters et al., 2013). Such variability also leads to increased energetic demands for thermoregulatory processes, as highlighted in Chapter 2. Salama et al. (2016) discuss the effects of thermal stress in ruminants in the face of future heat and cold waves of greater frequency, intensity, and duration. As an example, heat stress reduces feed intake in ruminants while increasing maintenance requirements. This challenge is compounded by a predicted decrease in the quality and productivity of feed resources available to herbivores grazing in rangelands (Craine et al., 2012). All these effects will inevitably lead to reductions in animal welfare and production. The research and managerial avenues Salama et al. (2016) suggest for reducing the negative impacts of climate variability on range animals involve improved monitoring methods that will allow managers to better assess thermal stress followed by innovative feeding practices and environmental modifications as well as improvements in thermal tolerance through more research on animal genetics.

Linked to the challenge of thermal stress is the problem of water availability in rangelands. Human population in rangelands is predicted to double by 2050. This increment will impose significant constraints on the already limited water supply of these typically dry regions with substantial implications for rangeland ecosystems (Vavra and Brown, 2006; Thornton et al., 2009) and in turn for the health and welfare of animal populations. Revell (2016) in Chapter 3 describes the welfare and production implications for rangeland livestock in a world with limited water availability. Heat stress inevitably interacts with water availability as water is essential in evaporative cooling mechanisms. On the other hand, low temperatures may reduce access to water which may inhibit food intake and, as a consequence, energy needed for thermoregulation. As for thermal stress, animal genetics may represent an important avenue for improving animal tolerance to a reduced water supply. For instance, a heritage breed from northern Mexico, the criollo cattle, is adapted to conditions of low and variable rainfall and hot temperatures and dry conditions. These animals tend to forage across a much larger area and away from water relative to Angus breeds (Peinetti et al., 2011). Thus, resorting to breeds already adapted to harsh conditions may be one approach to attenuate the consequences of heat stress and reduced water supply predicted for the future. Alternatively, as described by Revell (2016) in Chapter 3, early-life programming (i.e., through epigenetic effects) may represent another intervention aimed at improving the regulation of water balance in livestock. In addition to animal genetics and epigenetics, environmental experiences through behaviour-based management practices may also represent an alternative to increase the breadth of positive experiences that livestock could potentially gain in diverse and dry rangeland landscapes.

Water is a key resource fundamental to life, and as discussed by Bailey (2016) in Chapter 4, it represents the number one concern for ranchers and land managers when referring to animal distribution in extensive systems. Chapter 4 focuses on this fundamental variable which separates range animals from those being fed in confinement. Animal distribution is not only important for the nutrition and welfare of animals but also for the integrity and sustainability of rangeland ecosystems, which ultimately underlies the nutrition, health, and welfare of range animals. Animal distribution is influenced by biotic and abiotic factors inherent in a certain landscape as well as by managerial interventions which influence stocking rate. Stocking rate is one of the most important factors affecting forage availability and quality in rangelands (Edwards, 1980; Allison, 1985; De Villiers et al., 1994). Control of stocking rates is not only important for current generations of animals but also for future generations. Range degradation due to heavy stocking rates reduce carrying capacity and consequently the welfare of animals. In addition, high stocking rates can exacerbate the effects of drought on vegetation by increasing mortality of perennial grasses, negatively impacting on soils and thus contributing to range degradation and weed invasion (Stonecipher, 2015). On the other hand, conservative stocking rates lead to underuse of forage in wet years which may be inefficient,

particularly as environmental conditions (e.g., rainfall across seasons) become more variable. This is why it has been proposed that fixed and constant stocking levels are flawed when applied to pastoral systems, because there is too much variation in temporal and spatial production of forage and because livestock are not generally kept in one place for the whole year (Bourn and Wint, 1994). A more dynamic approach entails a flexible strategy which matches stocking rates to forage supply and/or rainfall events (Grissom and Steffens, 2013), aiming at securing quality and quantity of forage and, as a consequence, enhanced animal welfare. Nevertheless, the inherent structure of grazing systems in different parts of the world – particularly in emerging countries – make such a dynamic approach difficult to implement. For instance, prevailing pastoralist systems in communal lands and opportunistic pastoral management in sub-Saharan Africa prevent an effective regulation of animal numbers (Bourn and Wint, 1994). As with water availability and thermal stress, Bailey (2016) highlights environmental adaptation and genetics as potential tools to mitigate the challenges that rangelands impose on animal distribution. Low-stress handling techniques represent another interesting alternative for improving the distribution of animals in the landscape as well as for enhancing their welfare (Bailey, 2016).

Herbivores grazing in rangelands not only face changes in ambient temperature, food, and water supply. In contrast to animals fed in confinement, they are faced with choices among plants which in addition to nutritional imbalances contain toxins. Pfister et al. (2016) in Chapter 5 describe the negative impacts that poisonous plants impose – chronically or acutely – on herbivores' physiological processes and as a consequence on animal welfare. Availability of alternatives may reduce the negative impacts of poisonous plants on livestock because toxin–toxin interactions or interaction of toxins with an array of different chemicals present in a diverse diet may attenuate the negative impacts of single toxins (Provenza and Villalba, 2006). Range degradation and reductions in carrying capacity due to weed invasions decrease the availability of forage alternatives to grazing ruminants and, as a consequence, increase the likelihood of poisoning (Stonecipher, 2015). In addition to these changes, elevated atmospheric CO_2 concentrations and changes in ambient temperature may influence the concentration of toxins in some plants as well as their size and density (Pfister et al., 2016, Chaper 5). These effects could be compounded with future reductions in forage quality and abundance due to climate variability (Craine et al., 2012) and a decreased consumer tolerance to toxins under a warmer climate (Dearing, 2013). All these predicted effects suggest new challenges for welfare of animals grazing in rangelands. Pfister et al. (2016) comment on different management practices developed at their labs focused on reducing the likelihood of consumption of poisonous plants by livestock, like the induction of conditioned food aversion using the emetic drug lithium chloride (Provenza, 1995a, b; Ralphs and Provenza, 1999). As in previous chapters, Pfister et al. (2016) stress the importance of exploring phenotypic and genetic adaptations as tools to enhance the health and welfare of animals challenged by poisonous plants.

In this regard, genetic markers appear as a promising option for making important management decisions about grazing resistant or susceptible animals in a particular toxic plant environment (Scholtz et al., 2013).

Predation, Health, Neonatal Survival, and Transport

Laundré (2016) in Chapter 6 puts forward an analysis suggesting that livestock losses due to predators worldwide are relatively low. He suggests and gives examples where non-lethal management and policy methods may lead to healthy coexistence between livestock and predators. This coexistence could be beneficial as it allows for the maintenance of the positive roles that predators play in natural ecosystems such as the control of native ungulate numbers and, as a consequence, prevention of habitat degradation. Predators may also encourage the movement of prey across the landscape which reduces habitat overuse and potential degradation of forage resources (Ripple et al., 2014). In the case that livestock losses to predators could be reduced even further and if we put forward a scenario where coexistence is possible, what about the non-lethal effects of predators on their prey? We could, for instance, think about stress levels in animals that have evolved under predation pressure. The response of wild ungulates to predator presence appears to impact their feeding behaviour (e.g., increased time invested in vigilance), nutritional status (e.g., movement to poorer feeding sites due to predation risk), and social structure, all with negative impacts on animal welfare (Laundré et al., 2001; Hernández and Laundré,

2005). Laundré (2016; Chapter 6) points out that it is likely that domestic livestock responds in a similar way, although information available on the topic is scarcer and sometimes confounded by different managerial practices at different locations. Given the novelty of the topic, particularly for livestock, multiple avenues for research are open as several knowledge gaps need to be filled. For instance, whether the presence of non-lethal methods of predator control (e.g., guard dogs) may lead to reduced levels of stress in ungulates is not yet clear.

Grazing animals in rangelands are exposed to different health challenges from those housed in more intensive systems. Goddard (2016) in Chapter 7 explores these challenges rooted in managerial issues, patterns of exposure to infectious vectors, and environmental aspects. Health in range animals is being challenged by some of the same variables explored in previous chapters such as level of nutrition, water availability, and toxins. In addition, Goddard (2016) points out that climate warming may not only have its impact on plants and herbivores – as described in previous chapters – but also on the occurrence of a number of diseases through the increase in the geographic presence of infective vectors. As with previous challenges, monitoring, detection, and diagnosis of the problem are the first crucial steps for enhancing animal health. As a consequence, the presence of humans with the knowledge and training to perform such activities as well as the capacity for treatment is crucial. Intensive systems have a clear advantage in this regard. When looking into rangelands across the world there are instances where human–livestock interactions are only sporadic

and seasonal. In such conditions, new remote sensing methodologies are gaining traction for monitoring and management of grazing livestock production systems in rangelands (Gonzalez et al., 2014). These approaches will inevitably increase the already infrequent interactions between livestock and humans. On the other side of the spectrum reside systems where human–livestock interactions are very frequent, to a similar degree or even greater than those found in intensive systems. For instance, there are pastoral systems which rely on livestock movement and herding (e.g., Herrero et al., 2009) or systems where humans keep their livestock in their households for non-income and socio-cultural functions (e.g., Ouma et al., 2007). This close contact with livestock does not necessarily imply improved nutrition, welfare, or health for animals kept in such systems as several studies have shown that technical, institutional, and infrastructural constraints typically lead to the opposite outcome (Tano et al. 2003; Ouma et al., 2007). Goddard (2016) suggests that the most significant ways that extensive systems differ from intensive ones relate to the inability to rapidly identify a disease and then treat it. These issues are compounded by the increasing prevalence of microorganism resistance as a consequence of blanket chemotherapy (Geertz et al., 1997; Waller, 2006) and the use of diluted concentration of bioactive products (Plotkin, 2000) which complicate treatment. Finally, Goddard (2016) suggests that targeted treatments, remote sensing, and digital technologies to identify and diagnose disease are emerging approaches being successfully used in intensive systems which could potentially be implemented in extensive systems to improve the health and welfare of range animals.

The transition from the safety and constancy of the amniotic fluid to the external environment is a critical period for mammals and represents a serious welfare issue for animals kept in extensive systems. Dwyer and Baxter (2016; Chapter 8) describe the potential risks of neonatal mortality in extensive systems as well as opportunities to improve survival and to enhance the welfare of mothers and offspring living in these systems. In addition to the stark contrast between conditions in the uterus vs. the outside world, the neonate faces several of the challenges described in previous chapters such as threats from predators, lack of adequate nutrition and extremes of temperature. Given this scenario, Dwyer and Baxter (2016) suggest that the behavioural abilities of mother and young in rangelands become more important than those observed in intensive systems in order to secure neonate survival and to enhance welfare. In fact, the main causes of neonatal survival in extensive systems are tied to appropriate maternal care and offspring behavioural responses. As described above, in several nomadic and non-nomadic production systems worldwide there is close and daily interaction between humans and livestock. However, extensive systems in general reduce the likelihood of human interventions during birth. This characteristic may represent an advantage for species sensitive to disturbance, provided that interventions are not required to increase the survival capabilities of the offspring. Dwyer and Baxter (2016) also suggest that, as described in Chapter 2, the provision of appropriate thermal cover will improve welfare and enhance neonatal survival. As in previous chapters, genetic selection represents another avenue for improving welfare and neonatal survival, particularly the expression of desirable maternal and neonatal behaviours.

Humans transport animals to different distances with different purposes within multiple production systems worldwide. Distances and types of transportation range from long journeys across continents in sea vessels or airplanes to short distances in small vehicles. Phillips (2016) in Chapter 9 describes the transport of live animals and the welfare issues surrounding such a wide variety of journeys and distances. Several of the challenges faced by animals during transport are similar to those described in previous chapters: infectious diseases, thermal stress, thirst, and hunger. Additional challenges include vessel movement (leading to motion sickness and injury) and inefficient ventilation systems, which may lead to ammonia accumulation and other negative consequences to animal welfare. During transport, animals are prone to stress due to unfamiliarity with the environment, congregation and mixing with other groups, and to more frequent contact with humans. In addition to these variables, animals are gathered and handled prior to transport through different methods which could potentially increase the stress levels experienced during transport. Phillips (2016) concludes that more comparative research is needed to better understand animal welfare during different types and distances of transport and under the different conditions experienced during the process. The impacts of unfamiliarity with the environment as well as the effect of combinations of stressors on animal welfare are also identified as key areas for future research and understanding.

Rangelands are changing and this has a direct effect on the animals that inhabit these diverse and vast landscapes. Range animals are challenged by increases in habitat fragmentation, desertification and habitat loss, microbial infections, poisonous plants and predation, and by changes in climate and water supply. Challenges today may differ from those faced in the past but the adaptability and plasticity of animals to the problems imposed by their environment does not change. This adaptability comes about through genetic, epigenetic, and behavioural changes which allows for local adaptation (Provenza, 2008). We hope that this book stimulates new philosophies, ideas, research, and management approaches that enhance the ability of animals to adapt to the current and predicted harsh conditions imposed by rangelands. Such approaches should aim for adaptations and coping strategies that do not compromise animal welfare. Recommendations emerging from this book suggest that after a rapid and accurate diagnosis of the short- and long-term implications of the problem, this key objective could be achieved mainly by two means: 1) Making the animal better fit its environment through genotypic (e.g., new and heritage breeds) and epigenetic adaptations (e.g., locally-adapted animals); and 2) Continue looking at the optimal means to modify the environment (e.g., water supply, cover, supplemental feed, low-stress managerial practices, and guard dogs) in ways that better fit the animal. In conclusion, focusing on the adaptability and flexibility of animals to current and new environmental challenges will lead to the enhanced welfare of animals living in extensive systems.

REFERENCES

Allison, C.D. (1985) Factors affecting forage intake by range ruminants: A review. *Journal of Range Management* 38, 305–311.

Bourn, D., and Wint, W. (1994) *Livestock, land use and agricultural intensification in sub-Saharan Africa* (pp. 1–24). Overseas Development Institute, Pastoral Development Network.

Broom, D.M. (1986) Indicators of poor welfare. *British Veterinary Journal* 142, 524–526.

Craine, J.M., Nippert, J.B, Elmore, A.J., Skibbe, A.M., Hutchinson, S.L., and Brunsell, N.A. (2012) Timing of climate variability and grassland productivity. *Proceedings of the National Academy of Sciences* 109(9), 3401–3405.

Dearing, M.D. (2013) Temperature-dependent toxicity in mammals 474 with implications for herbivores: A review. *Journal of Comparative Physiology B* 183, 43–50.

De Villiers, J.F., Botha, W.A., and Wandrag, J.J. (1994) The performance of lambs on kikuyu as influenced by stocking rate and grazing system. *South African Journal of Animal Science* 24, 133–139.

Duncan, I.J.H., and Fraser, D. (1997) Understanding animal welfare. In: Appleby, M.C., Hughes, B.O. (Eds.) *Animal Welfare*. CAB International, Wallingford. pp. 19–31.

Edwards, P.J. (1980) The use of stocking rate/animal performance models in research and extension. Proceedings of the Grassland Society of South Africa 15, 73–77.

FAO (2011) The state of the world's land and water resources for food and agriculture (SOLAW)—Managing systems at risk. Food and Agriculture Organization of the United Nations, Rome, Italy.

Farm Animal Welfare Council (1992) FAWC updates the Five Freedoms. *Veterinary Record* 17, 357.

Geertz S., Coles G.C., and Gryseels, B. (1997) Anthelmintic resistance in human helminths: Learning from the problems with worm control in livestock. *Parasitology Today* 13, 149–151.

González, L.A., Bishop-Hurley, G., Henry, D., and Charmley, E. (2014) Wireless sensor networks to study, monitor and manage cattle in grazing systems. *Animal Production Science* 54(10), 1687–1693.

Grissom, G., and Steffens, T. (2013) Case study: adaptive grazing management at Rancho Largo Cattle Company. *Rangelands* 35(5), 35–44.

Hernández, L., and Laundré, J.W. (2005) Foraging in the 'landscape of fear' and its implications for habitat use and diet quality of elk *Cervus elaphus* and bison *Bison bison. Wildlife Biology* 11, 215–220.

Herrero, M., Thornton, P.K., Gerber, P., and Reid, R.S. (2009) Livestock, livelihoods and the environment: understanding the trade-offs. *Current Opinion in Environmental Sustainability* 1(2), 111–120.

IPCC (2013) Climate change 2013: The physical science basis. Fifth assessment report of the Intergovernmental Panel on Climate Change. Cambridge University Press, Cambridge, UK and New York, NY, USA.

Laundré, J.W., Hernández, L., and Altendorf, K.B. (2001) Wolves, elk, and bison: reestablishing the "landscape of fear" in Yellowstone National Park, U.S.A. *Canadian Journal of Zoology* 79, 1401–1409.

Lovina, R., Launchbaugh, K., Jones, J., Babcock, L., Ambrosek, R., Stebleton, A., Brewer, T., Sanders, K., Mink, J., and Hyde G. (2009) *Rangelands: An introduction to Idaho's wild open spaces.* Department of Rangeland Ecology and Management, University of Idaho, Moscow, Idaho and Idaho Rangeland Resource Commission, Emmett, Idaho.

Ouma, E., Abdulai, A., and Drucker, A. (2007) Measuring heterogeneous preferences for cattle traits among cattle-keeping households in East Africa. *American Journal of Agricultural Economics* 89(4), 1005–1019.

Peinetti, H.R., Fredrickson, E.L., Peters, D.P., Cibils, A.F., Roacho-Estrada, J.O., and Laliberte, A.S. (2011) Foraging behavior of heritage versus recently introduced herbivores on desert landscapes of the American Southwest. *Ecosphere* 2(5), Article 57, 1–14.

Peters, D., Archer, S., Bestelmeyer, B., Brooks, M., Brown, J., Comrie, A., Gimblett, H.R., López-Hoffman, L., Sala, O.E., Vivoni, E., and Brooks, M.L. (2013) Desertification of rangelands. In: Pielke, R. (Ed.) *Vulnerability of ecosystems to climate.*, Academic Press, Oxford, UK, pp. 239–258. doi:10.1016/B978-0-12-384703-4.00426-3

Plotkin, M.J. (2000) *Medicine quest. In search of Nature's healing secrets.* Penguin Putnam Inc., New York, NY, USA.

Provenza, F.D. (1995a) Postingestive feedback as an elementary determinant of food preference and intake in ruminants. *Journal of Range Management* 48, 2–17.

Provenza, F.D. (1995b) Tracking variable environments: there is more than one kind of memory. *Journal of Chemical Ecology* 21, 911–923.

Provenza, F.D. (2008). What does it mean to be locally adapted and who cares anyway? *Journal of Animal Science* 86(14): E271-E284.

Provenza, F.D., and Villalba J.J. (2006) Foraging in domestic herbivores: Linking the internal and external milieu. In: Bels, V.L. (Ed.) *Feeding in domestic vertebrates: From structure to function.* CABI Publ., Oxfordshire, UK. pp. 210–240.

Ralphs, M.H., and Provenza, F.D. (1999) Conditioned food aversions: principles and practices, with special reference to social facilitation. *Proceedings of the Nutrition Society* 58, 813–820.

Ripple, W.J., Estes, J.A., Beschta, R.L., Wilmers, C.C., Ritchie, E.G., Hebblewhite,

M., Berger, J., Elmhagen, B., Letric, M., Nelson, M.P., Schmitz, O.J., Smith, D.W., Wallach, A.D., and Wirsing, A.J. (2014) Status and ecological effects of the World's largest carnivores. *Science* 343, 152–162.

Scholtz, M., Maiwashe, A., Neser, F., Theunissen, A., Olivier, W., Mokolobate, M., and Hendriks, J. (2013) Livestock breeding for sustainability to mitigate global warming, with the emphasis on developing countries. *South African Journal of Animal Science* 43: 269–281.

Society for Range Management. http://www.rangelands.org/

Stonecipher, C. (2015) Mitigation of medusahead and lupine induced crooked calf syndrome through grazing and revegetation on the channel scablands of eastern Washington. PhD Dissertation. Utha State University, Logan, Utah, USA.

Tano, K., Kamuanga M., Faminow M., and Swallow B. (2003) Using conjoint analysis to estimate farmer's preferences for cattle traits in West Africa. *Ecological Economics* 45, 393–407.

Thornton, P.K., Van de Steeg, J., Notenbaert, A., and Herrero, M. (2009) The impacts of climate change on livestock and livestock systems in developing countries: A review of what we know and what we need to know. *Agricultural Systems* 101(3), 113–127.

Van Soest, P.J. (1994) *Nutritional ecology of the ruminant*. Cornell University Press, Ithaca, New York.

Vavra, M., and Brown, J. (2006) Rangeland research: strategies for providing sustainability and stewardship to the rangelands of the world. *Rangelands* 28(6), 7–14.

Villalba, J.J., Manteca, X.P., Vercoe, E. Maloney, S.K., and Blache, D. (2016) Integrating nutrition and animal welfare in extensive systems. In: Phillips, C. (Ed.) *Nutrition and the Welfare of Farm Animals*. Springer, Heidelberg, New York. In press.

Waller P.J. (2006) Sustainable nematode parasite control strategies for ruminant livestock by grazing management and biological control. *Animal Feed Science and Technology* 126: 277–289.

chapter two

Thermal stress in ruminants: Responses and strategies for alleviation

Ahmed A.K. Salama[1,2], Gerardo Caja[1], Soufiane Hamzaoui[1], Xavier Such[1], Elena Albanell[1], Bouabid Badaoui[3], and Juan J. Loor[4]

INTRODUCTION

Searching the PubMed database for the terms "heat stress" and "cold stress" resulted in 8712 and 3181 articles, respectively, listed up to September 28, 2015. The number of articles dealing with both heat and cold stress increased exponentially from 20 in 1970 to 968 in 2014 on average, and is expected to continue increasing in the future. This high impact on the scientific literature reflects a significant concern about the environmental consequences of extremes in global temperatures. According to the most recent assessment report of the Intergovernmental Panel on Climate Change (IPCC, 2013), the observed earth's surface temperature increased by up to 2.5°C from 1901 to 2012. Furthermore, in the Northern hemisphere, 1983–2012 was the warmest 30-year period of the last 1400 years. Continued emissions of greenhouse gases will cause further warming. By the end of the twenty-first century, the models of IPCC projected a 3–10°C warming in annual temperature over the global land area, with greater warming in the Northern hemisphere. Taking temperature and precipitation projections together, the climate would shift toward warmer and drier climate types (Feng et al., 2014). For grazing livestock production systems, the effect of climate change will be more noticeable due to the fact that drought will affect negatively vegetation growth in pastures jointly with the direct effects of high temperature and solar radiation on livestock (Silanikove and Koluman, 2015).

The cold stress, on the other hand, is apparently less important for bovine (dairy and beef), sheep, and goats. This is because these animals were claimed

[1] Grup de Recerca en Remugants (G2R), Departament de Ciència Animal i dels Aliments, Universitat Autònoma de Barcelona, Spain
[2] Sheep and Goat Research Department, Animal Production Research Institute, Giza, Egypt
[3] Science Faculty, Mohamed V University, Rabat, Morocco
[4] Department of Animal Sciences and Division of Nutritional Sciences, University of Illinois, Urbana-Champaign, USA

to have a low critical temperature (−35 to −20 °C) that rarely occur in agricultural regions. Despite this theoretical cold tolerance of ruminants, productive data clearly indicate a marked seasonal fluctuation at temperatures well above these reported lower critical temperatures. This is due to the fact that reported critical lower temperatures came from short-term laboratory studies with still cool air and may be somewhat irrelevant when one considers physiological changes and factors affecting animal production rather than the short-term calorimetric responses.

In the future, it is likely that farm animals under intensive as well as extensive production systems will face heat and cold waves with a higher frequency, intensity, and duration, which will affect their health, welfare, and performance. This is a complicated issue, especially for countries where the production system has a high diversity (species, breeds, nutrition) because it means that different approaches will be needed according to the situation to reduce the impact of heat stress. Furthermore, some regions have extremely low ambient temperatures in winter and high temperature records in summer, which represents a challenge for animals in these regions to cope with both heat- and cold-stress.

Comfortable animals are productive and healthy animals. Although a significant effort has been placed on research related to the impact of thermal stress on livestock animals, many points still need to be clarified or resolved. The high variability between species, breeds, and even among individual animals within a breed complicates the adoption of a specific mitigation strategy. Throughout this chapter, the reader will realize that there are numerous excellent and comprehensive reviews describing the effects of heat and cold stresses as well as strategies to mitigate their adverse effects (Fuquay, 1981; Young, 1981, 1983; Collier et al., 1982, 2006, 2008; Silanikove, 2000a, b; Kadzere et al., 2002; West, 1999, 2003; Hansen, 2004; Marai et al., 2007; Bernabucci et al., 2010; Marai and Haeeb, 2010; Baumgard and Rhoads, 2012, 2013; Sevi and Caroprese, 2012; Rhoads et al., 2013, Mader, 2014; Salama et al., 2014). However, most of these review articles are on dairy animals in intensive systems and just a few deal with beef cattle and other ruminants in extensive systems. The aim of the present chapter is to present an overview of the current state of knowledge on thermal stress, focusing on differences between species for their responses to both heat and cold stress.

THERMOREGULATION IN RUMINANTS: INTERACTION BETWEEN ENVIRONMENTAL CONDITIONS AND ANIMAL CHARACTERISTICS

All homeothermic animals, including ruminants, need to maintain their body temperature within a very narrow interval regardless of heat production and the environmental conditions. Maintaining the body temperature is necessary for optimal physiological functions and cellular metabolic reactions. According to Lee (1965), heat load on the animal is affected by environmental variables (temperature, humidity, air movement, radiation, precipitation) and animal characteristics (species, breed, sex, age, physiological state, coat, acclimatization, nutrition, health status). These many factors make it complex to describe and predict the impact of environment on dairy ruminants. With similar body size and surface area, the lactating cow has significantly more heat to dissipate than a dry cow and will have greater difficulty dissipating the heat during hot and humid conditions. Purwanto et al. (1990) reported that low (19 kg/d) and high (32 kg/d) yielding cows generated 27 and 48% more heat than non-lactating cows despite having lower body weight. One of the nutritional factors that could affect the thermoregulation in grazing animals is the fact that some forages, such as tall fescue and ryegrass, can be endophyte-infected, which produces alkaloids that raise body temperature, making the thermoregulation difficult under high ambient temperatures.

Approximately one third of energy intake is transformed to heat production in dairy cows (Coppock, 1985). The thermoneutral heat production of beef cattle ranges from about 100 W/m^2 at maintenance to about 160 W/m^2 when consuming a production ration ad libitum (Webster, 1974). The heat accumulated by the animal (heat gain) is the sum of heat accumulated from the environment and the heat associated with metabolic processes. In this regard, under heat stress (HS) conditions, the animal should lose heat to maintain its body temperature. It is well known that outside the thermoneutral zone, the animal is cold- (below the lower critical temperature) or heat-stressed (above the upper critical temperature), and extra energy for maintenance is necessary in both cases. For instance, under HS there are extra maintenance expenditures due to muscle movements for panting, N losses

with sweating, increased chemical reactions in the body and the production of heat shock proteins that consume large amounts of ATP. In parallel with the increase in maintenance requirements, there is a clear reduction in feed intake, which is negatively reflected on animal productivity. On the other hand, below the lower critical temperature (cold stress) the animal uses an extra fraction of energy to keep warm through shivering and other heat-producing processes. According to NRC (2007), energy requirements for maintenance are 20% greater in cold winters, and if an animal's body surface is wet and not protected from the wind, then these requirements can easily double. Animals tolerant to HS have the ability to reduce heat production, increasing heat loss to the environment, or a combination of both of these physiological processes. On the other hand, animals are more tolerant to cold stress (CS) when they are able to increase heat production, decrease heat loss, or a combination of both. Consequently, selection for greater HS tolerance could result in less tolerant animals to CS and vice versa.

Heat can be dissipated by sensible (radiation, conduction, convection) and latent (evaporation) mechanisms. The three sensible ways could also be sources to gain heat rather than losing it, especially when the ambient temperature is high or the animal is directly exposed to solar radiation. In these conditions of high heat load, heat dissipation is shifted from radiation, conduction, and convection to evaporation (sweating and panting). In cold environments an animal attempts to minimize evaporation to reduce heat loss. The heat exchange rate is clearly affected by ambient temperature, but could also be modified by the animal's behavior (e.g., shelter seeking, postural changes) and by changing its thermal insulation through retention or shedding of hair, hair piloerection, or the control of blood flow to superficial tissue.

Coat color and thickness, and hair density are among the animal factors that affect thermoregulation. Cows have apocrine sweat glands and one sweat gland associated with each hair fiber. Consequently, hair density directly reflects the number of sweat glands, and hair diameter and length have effects on evaporative heat loss by regulating airflow at the skin surface (Collier et al., 2008). Collier et al. (2008) reported the existence of a single gene affecting hair coat density and hair length. Cows bearing that gene have a greater ability to deliver heat from the body core to the skin (evaporative heat loss) under HS conditions (see later).

Compared to dairy livestock, beef animals are often managed under extensive conditions with minimal possibilities for environmental modifications, making body temperature regulation a challenge for the animal. There is a clear difference between beef breeds in their thermoregulation and in their immune responses under thermal stress conditions. As shown by Carroll et al. (2012), Angus heifers displayed greater rectal temperature than Colombian heat-tolerant Romosinuano heifers when maintained under either thermal neutral or HS conditions and produced a greater overall febrile response to lipopolysaccharide challenge (i.e., had greater average rectal temperature after challenge).

In water buffaloes the skin is covered with a thick epidermis, containing many melanin particles that give the black color and trap ultraviolet rays, preventing them from penetrating the skin to the subcutaneous tissue (Marai and Haeeb, 2010). Buffaloes also have scarce hair and the skin sebaceous glands are very active compared to cows. These glands secrete greasy material that is melted and becomes glossy under the sun, reflecting sunrays and decreasing the heat load by radiation. However, buffalo skin has a lower density of sweat glands than cows, so buffaloes dissipate heat poorly by evaporation (Marai and Haeeb, 2010), and they usually look for protection from the sun by staying in shade or resting in water.

Dromedary camels are well known for their high heat tolerance and they do not pant under severe HS. Besides their high capacity for sweating, camels are also able to dissipate a significant amount of heat by convection, where the vasodilation of peripheral vessels leads to an increase in cutaneous blood flow and heat dissipation (Abdoun et al., 2012). Long legs, sternal stair, and selective brain cooling are permanent camel adaptations to increase their ability to dissipate heat (Elkhawad, 1992; Souilem and Varosme, 2009).

Black goats are dominant in hot deserts and have advantages for solar radiation exposure over white goats (Finch et al., 1980). Although the black coat absorbs much more solar radiation, black goats are able to look for food for a longer time in the sun. Black goats drink a volume of water equal to about 35% of their body weight and are able to efficiently cool themselves by evaporation. Short haired goats exposed to solar radiation had greater increases in rectal and dermal temperatures, respiratory and pulse rates, and consumed less feed than the long haired goats (Acharya et al., 1995). Those authors concluded that long haired goats tolerated radiant heat better than short haired goats.

HOW TO MEASURE THERMAL STRESS LEVEL

Body temperature is an excellent indicator of the animal's susceptibility to thermal stress. However, devices used to monitor body temperature are usually not feasible for large numbers of animals under extensive conditions (Mader et al., 2006). The use of infrared thermography guns has been shown to be a low-cost approach to measure the skin surface temperature of animals. Additionally, the infrared thermographic cameras were also used as an indicator of heat production and skin temperature in cattle (Montanholi et al., 2008). Infrared skin temperature is highly correlated with respiration rates and is a good indicator of the microenvironment around the animal. Furthermore, the measurement can be taken from a distance, which does not require restricting movement of the animals.

Respiration rate is easier to assess under field conditions compared with rectal temperature and could be a good indicator of heat load, but is a time-consuming process. Panting score, on the other hand, is a function of respiration rate and body temperature, and could be easily visually recorded. Gaughan and Mader (2014) reported a strong relationship (R^2 = 0.83 to 0.94) between panting score (0–4.5; where 0 = no panting/no stress and 4.5 = extreme stress) and

body temperature in un-shaded beef cattle.

Monitoring animals (e.g., body temperature or respiratory rate) is a key point to detect thermal stress and to take appropriate measures. In practical conditions it is not easy to measure body temperature and/or respiratory rate for each individual animal. In this case, measuring 10% of animals in the herd, or even a few animals, would be enough. However, it should be recognized that response to thermal stress could vary according to the production level (high productive animals are more sensitive) and there is the possibility of existence of microclimates in different locations. Besides measurements on animals, ambient temperature, humidity, and air velocity should also be recorded. The ThermalAid is an application for smartphones (University of Missouri, 2012) that uses live weather data and cattle respiration rates to monitor the degree of HS and to provide tips to minimize the effects of heat load.

As mentioned above, measurements on animals (e.g., body temperature and respiration rate) are not easy. Therefore, a temperature humidity index (THI) has been proposed as an alternative to determine the extent to which the animal is heat-stressed. Dikmen and Hansen (2009) reported eight equations for the calculation of THI. One common equation to calculate THI is: THI = (1.8 × ambient temperature + 32) − [(0.55 − 0.0055 × relative humidity) × (1.8 × ambient temperature − 26.8)]. Generally, beef cattle start to suffer thermal stress at THI outside the range of 35 to 74 (Gaughan and Mader, 2009), which corresponds to −3 and +27°C at 50% relative humidity, respectively. For dairy cows, Dikmen and Hansen (2009) reported that the upper critical THI (the THI at which rectal temperature is 38.5°C) was 78, whereas Bohmanova et al. (2007) reported a threshold of 72 for cows in Georgia and 74 for cows in Arizona. This variation in the upper critical THI value could be due to differences in the adaptation of cows to HS and/or other features related to their housing conditions. The upper critical THI is not well defined in case of dairy goats, but from our observations it could be the THI value causing rectal temperature above 39.0°C.

The THI indices only take into account the temperature and relative humidity. Mader et al. (2006) proposed adjustments to the THI for solar radiation (W/m^2) and wind speed (m/s) because solar radiation can greatly influence heat load, whereas changes in wind speed result in altered convective cooling. In case of feedlot beef cattle, Mader et al. (2006) reported that THI values should be reduced by 2.0 units for each 1-m×s^{-1} increase in wind speed and increased by 0.68 units for each 100-W×m^{-2} increase in solar radiation. Furthermore, Bryant et al. (2010) adjusted the THI index by adding wind speed (m/s), solar radiation (MJ/m^2), and the minimum temperature in a heat load index developed in New Zealand. These adjustments could be useful with grazing animals under extensive production systems, and especially for dark-coated animals.

Indices for CS are less studied than those for HS in livestock animals. Sipple and Passel (1945) proposed a wind-chill index (WCI) for humans that relates ambient temperature and wind speed to the time for freezing water (using a plastic bag full of water at different wind speeds during cold to determine how fast the temperature in water would drop). This index could be suitable to predict the cooling power of cold and wind combinations for bare-skinned animals, but may not be valid for animals with hair or wool. Later, Ames and Insley (1975) took into account the external insulation in cows (hair) and sheep (wool) and established a cubic equation that included the effect of wind speed on heat loss rate. They found that the rate of heat loss in cattle exposed to −10°C increased rapidly at wind speed from 0 to 16 km/h, then this increase was attenuated, and increased again at wind speeds greater than 40 km/h. Tew et al. (2002) developed a biologically-based equation for use in determining effects of wind (km/h) on humans as well as domestic livestock. This equation is as follow:

WCI = 13.112 + (0.6215 × ambient temperature) − (11.37 × wind speed $^{0.16}$) + (0.3965 × ambient temperature) × wind speed $^{0.16}$

More recently, based on data obtained from ranchers and scientific papers a cold advisory for newborn livestock (NOAA, 2009) was established.

ANIMAL WELFARE UNDER THERMAL STRESS: THE FIVE FREEDOMS

One of the current welfare concerns for livestock is weather extremes. Several regions in the world have experienced severe heat or cold stresses that have resulted in substantial animal mortality and significant economic losses. Consequently, the livestock industry must utilize management tools that help farmers and ranchers to monitor animal responses to thermal stress and minimize the negative effects of these adverse climatic conditions.

Freedom from Thirst, Hunger, and Malnutrition

Water availability is essential under thermal stress situations. As shown by Revell (2016; see Chapter 3) the demand for water increases in hot conditions. Under HS, the animal gets rid of the heat load by evaporation (respiration and sweating), which increases the need for drinking water. Besides its use for evaporative cooling, the high specific heat of water allows the animal to absorb a large amount of heat during the day and dissipate it during the night if cool. Therefore, a source of available water is essential in high ambient temperatures. Thompson (1985) estimated the heat loss in sheep to approximately 20% of total body heat via respiratory moisture in a neutral environmental temperature (12°C). However, the moisture loss increases and accounts for approximately 60% of the total heat loss at high ambient temperature (35°C). Sweating rate in dairy cows increased from 29.6 g/m^2 per hour at 18°C to 67.5 g/m^2 per hour (+228%) at 33°C (Shwartz et al., 2009). The expected climate change and the increase in ambient temperatures will lead to the shrinkage of water resources, which represents a significant threat for HS animals that need greater amounts of water under such conditions. In very cold temperatures, if water is frozen, animals will stop drinking, reduce food consumption, and will not be able to produce extra metabolic heat necessary to maintain body temperature.

As mentioned above, in high temperatures animals need more water to cool themselves by evaporation (increased respiratory rate and sweating). Dry dairy cows at 30°C were able to keep similar rectal temperature (38.5°C) to cows maintained at 20°C by increasing respiratory rate from 18 to 38 breaths/min (Itoh et al., 1998). However, if HS persists, the respiratory rate will dramatically increase, but the evaporative mechanism alone would not be enough to dissipate the body heat. Consequently, body temperature will increase (39.1 to 41.0°C or greater) with a consequent decrease in food consumption. Reducing feed intake is a kind of spontaneous thermoregulatory behavior to decrease heat production. In extensive production systems where animals are grazing, besides the high ambient temperature in summer there is a clear reduction in the forage quality (high fiber and lower protein contents), which negatively affects the animal welfare.

Freedom from Discomfort

Beef cattle are most comfortable when temperatures range between 5 and 20°C. As the temperature rises above this level, cattle slowly begin showing signs of HS as mentioned previously. On hot days, it is easy to detect animals with open mouth and saliva drooling (Schütz et al., 2014). Heat-stressed animals are not comfortable and move a lot seeking cooler zones. In addition, below the lower critical temperature, the acute response of animals is shivering to increase heat production. Cold stress has a marked impact if the amount and/or quality of food is not adequate because animals need extra-energy consumption for heat production. Therefore, it is necessary that caretakers guarantee suitable shelter according to the season (i.e., shade during hot summer and shelter or windbreaks during winter, as shown later).

Freedom from Pain, Injury, and Disease

Thermal stress compromises the metabolic state of the animal and consequently its health. Heat-stressed animals are more prone to metabolic alkalosis and subacute rumen acidosis. Although HS cows eat less food (negative energy balance), they do not use body fat reserves, but they mobilize body protein (muscle degradation) and use amino acids as an energy source for glucose production (Baumgard and Rhoads, 2013). We detected a greater level of blood creatinine as a muscle degradation indicator in HS goats (Salama et al., 2014), and blood creatinine correlated positively with blood urea-N ($r = 0.41$; $P < 0.05$) and blood glucose ($r = 0.33$; $P < 0.10$) as shown in Table 2.1. Due to muscle

degradation, HS animals suffer significant body weight losses.

Prolonged exposure to HS alters the acid-base status due to changes in the relationship between anions and cations in the blood. Furthermore, panting decreases blood CO_2 via pulmonary ventilation, reducing the blood concentration of carbonic acid, and resulting in a respiratory alkalosis (greater blood pH). Compensation for the respiratory alkalosis involves increased urinary HCO_3^- excretion, which leads to a decline in blood HCO_3^- concentration. There was an increase in blood Cl^- of HS goats (Hamzaoui et al., 2013) and dairy cows (Calamari et al., 2007). The greater clearance of blood HCO_3^- (by its secretion in urine by kidneys) is related to the increase in blood Cl^- as there is Cl^-/HCO_3^- exchanger in kidney cells (Verlander, 1997). Accordingly, we detected a negative correlation ($P < 0.01$) between blood HCO_3^- and Cl^- in thermoneutral ($r = -0.46$) and HS ($r = -0.51$) goats. However, in intensive heat stress

Table 2.1 Correlations among different basal levels of blood metabolites in goats under thermal neutral (above diagonal) and heat stress (below diagonal) conditions

	CREA	BUN	GLU	NEFA	BHB	LAC	INS
CREA		0.30	-0.08	-0.02	0.11	-0.18	-0.33
BUN	0.41*		0.18	0.25	0.21	0.13	-0.23
GLU	0.33f	0.07		-0.15	-0.21	0.32f	0.39*
NEFA	0.04	0.11	-0.38*		-0.10	0.20	-0.19
BHB	-0.08	-0.08	-0.47*	0.29*		-0.08	-0.07
LAC	0.16	0.19	0.11	0.21	-0.33*		0.35
INS	0.16	0.53*	0.37*	-0.28	-0.23	0.13	

f $P < 0.10$

* $P < 0.05$

Abbreviations: CREA = creatinine, BUN = blood urea N, GLU = glucose, NEFA = non-esterified fatty acids, BHB = β- hydroxybutyrate, LAC = lactate, and INS = insulin.

there is overcompensation for alkalosis during the cooler part of the night, and consequently animals could suffer metabolic acidosis rather than alkalosis.

HS affects both pH and temperature of the rumen, and consequently could alter rumen microbial populations and make the animal more prone to subacute ruminal acidosis. Recently, we used wireless rumen boluses to monitor changes in ruminal pH and temperature in dry goats fed at maintenance level under thermoneutral and HS conditions (Castro-Costa et al., 2015). Rumen and rectal temperatures correlated positively (R^2 = 0.62; $P < 0.01$), the rumen temperature being greater than the rectal temperature by +0.95 ± 0.11°C. Despite eating the same amounts of dry matter (DM), mean daily rumen pH of the HS goats was 0.12 units lower than that of the thermoneutral goats. Similarly, Mishra et al. (1970) and Bandaranayaka and Holmes (1976) reported that rumen pH in HS cows fed a similar amount of DM was lower than controls. The possible reasons for the lower pH under HS conditions are:

- Eating less with shorter rumination time (*lower saliva production*).
- Greater excretion of HCO_3^- in urine to compensate for the greater exhaled CO_2 during panting, and, thus, less HCO_3^- content of saliva (*lower saliva quality*).
- Greater blood flow to the periphery (for heat dissipation) and reduced blood flow to the gastrointestinal tract, which reduces the absorption efficiency (*volatile fatty acid accumulation in the rumen*).
- Reduction in eating frequency, but with larger meals (*more fermentation post-eating*).

For the rumen temperature, the HS goats had, on average, greater values (+0.3°C) than the thermoneutral goats in accordance with the high ambient temperature under which the HS goats were housed (Castro-Costa et al., 2015). It is not clear to what extent these changes in ruminal pH (–0.12 units) and temperature (+0.3°C), induced by HS in dry goats, could affect rumen microflora, but we expect a greater effect in animals with greater feeding level. Profiling the rumen microbiota by 16s RNA gene cloning confirmed that HS induces significant changes in microbial diversity in heifers (Tajima et al., 2007).

During hypothermia in cold-stressed animals all metabolic and physiological processes begin to slow. Blood supply to the extremities is reduced (to protect vital organs), which could result in frostbite of the teats and ear tips. Rapid thawing of the tissue is preferable and less painful than gradual thawing. However, those cows that suffer frostbitten teats are more prone to mastitis.

Freedom to Express Normal Behavior

Domesticated ruminants are diurnal in nature, being active during the day and resting at night (Silanikove, 2000b). During hot weather, animals look for shade, change their orientation to the sun (decreasing exposure to solar load), and dramatically increase water intake (Hamzaoui et al., 2013). Additionally, grazing occurs at cooler times (before sunrise or during the night). These alterations in animal behavior help animals to minimize internal heat production. On the other hand, cold temperatures will increase heat loss, and animals will respond with physiological and behavioral mechanisms of thermoregulation to reduce heat loss (huddling) and increase heat production (shivering, increased DM intake). Wind and moisture (rain and snow) combined with low temperatures dramatically reduce the insulating value of an animal's hair coat and increases surface heat exchange, which will aggravate the negative impact of cold environment on the animal.

By using individual leg data loggers that measured position and intravaginal data loggers that measured core body temperature, Allen et al. (2015) reported that a body temperature of 38.9°C indicated a 50% likelihood a cow would be standing, providing the physiological evidence that standing may help cool. Nevertheless, this longer standing time was related to hoof lesions. When the THI increased from 56 to 74, lying time in dairy cows decreased by 28%, and the increases in claw horn lesion development reported in the late summer could be associated with an increase in total daily standing time (Cook et al., 2007). Overall, HS modifies the lying behavior and could result in leg problems.

Freedom from Fear and Distress

Animals in extensive systems are in limited contact with human and other factors that could cause fear and distress. However, animal handling at given times is necessary for management tasks (herding, vaccination, pregnancy detection, etc.), but could cause distress situations (Grandin, 1997). As mentioned above, monitoring is important for detecting thermal stress in animals and for taking measures to alleviate its effects. Inserting a thermometer in the rectum for core body temperature measurement or even approaching an animal for respiratory frequency counting could cause fear in the animal. Under these conditions, the use of infrared thermography guns and cameras could be good tools to measure skin surface temperature (Montanholi et al., 2008) and to predict the microenvironment around the animal. Additionally, the application of animal sensors could help in recording a large amount of behavioral and/or physiological data that could be wirelessly transmitted by readers located at water points when animals drink (Rutter, 2014). Thus, a high volume of information could be remotely collected without exposing animals to situations of fear or distress. For cost issues, it is enough that some representative animals in the herd (age, physiological state, etc.) wear the sensors.

IMPACT OF THERMAL STRESS ON ANIMAL PRODUCTIVITY

Heat stress reduces feed intake by 15 to 40% and at the same time increases maintenance requirements (+ 30%) necessary for the extra activities mentioned above (see 'Thermoregulation in ruminants'). Animals in extensive systems will stop grazing and seek shade (if possible), which dramatically decreases the dry matter intake. Consequently, energy intake would not be enough to cover the daily requirements for production (milk or meat). Furthermore, blood flow to the mammary gland (with the components needed for milk synthesis) is reduced (as blood flows to the skin for heat dissipation), which in turn reduces milk yield and milk components (Lough et al., 1990). A direct effect of HS on mammary cell synthetic capacity also is possible (see below, 'Gene expression regulation under thermal stress'). Baumgard and Rhoads (2013) indicated that reduced feed intake during HS only explained 50% of the decreased milk yield in dairy cows. In case of dairy goats, it seems that reduced feed intake could explain all of the milk yield losses under HS at mid- and late-lactation (Hamzaoui, 2014; Salama et al., 2014), but this assumption should be confirmed by pair-fed experiments in dairy goats. Similarly, in beef cattle O'Brien et al. (2010) indicated that HS caused a reduction in DM intake by 12% and this reduced feed intake during HS appeared to fully explain decreased average daily gain in bull calves.

Beef calves kept extensively in adverse winter conditions consume more feed, but grow slower (greater feed to gain ratios) because maintenance requirements are increased 30 to 70% (Young, 1981), and consequently less feed energy is available for productive processes. Generally, it is considered that for every 1°C drop below the lower critical temperature, there is an approximate 2% increase in energy requirements. If animals are not able to increase their feed intake (insufficient amount or low quality of forage), they will start to lose weight by mobilizing body fat and protein, consequently they will have less insulation and become more susceptible to cold stress. In lactating cows, low ambient temperatures reduce milk yield and increase milk fat percentages, with the effects most marked during the early stages of lactation. Appetite is also stimulated by cold and when feed is available, cows increase their intake, which may delay the depression in milk yield caused by cold. However, an increased feed consumption without an increase in yield will reduce feed efficiency.

There is a negative relationship between THI and milk production in dairy cows (Bernabucci et al., 2010), and for each unit increase in THI, there is a decrease in milk yield by 0.23 to 0.59 kg/d (Bohmanova et al., 2007). A difference between breeds exists in their tolerance to HS, which would allow dairy producers in climates prone to HS to select the most suitable cows for successful production. Smith et al. (2013) in their retrospective analyses reported that Jersey cows were more tolerant to hot conditions and suffered lower milk yield losses than Holstein cows. Furthermore, Jersey cows had greater productive efficiency (kg of energy corrected milk/kg of DM intake) than Holstein cows in commercial Danish herds (Kristensen et al., 2015). Greater tolerance to HS and better production efficiency might make Jersey cows the breed of choice under the expected challenge of climate change and its negative effects on dairy production.

For dairy ewes, Sevi and Caroprese (2012) reported a marked increase in rectal temperature, metabolism alteration, and a reduction of milk yield after ewe exposure, even for short periods, to average daily temperatures of 35°C or after prolonged ewe exposure to mean ambient temperatures of 30°C. Dairy goats exposed to THI 79 to 89 suffered milk yield losses by 3 to 13% (Sano et al., 1985; Brown et al., 1988; Hamzaoui et al., 2013). Furthermore, Salama et al. (2014) estimated that for each 1 unit increment of THI there is a decrease of 1% in milk yield. Although feed intake decreased similarly (−22 to −35%) in dairy goats and cows by environmental-induced HS, milk yield losses in dairy goats (–3 to –13%) were much lower than values reported in dairy cows (–27 to –33%) (Shwartz et al., 2009; Wheelock et al., 2010).

The HS results in greater diet digestibility in cows (McDowell et al., 1969), heifers (Bernabucci et al., 1999), and male goats (Hirayama et al., 2004). The increased digestibility under HS might be partly due to the reduction of feed intake. Another reason for the enhanced digestibility under HS conditions could be a reduced passage rate of solid digesta as reported by Bernabucci et al. (1999). We used ytterbium as a marker of solid digesta, and contrary to what was reported for heifers and cows, the passage rate was similar between thermoneutral and HS goats (Figure 2.1). On the other hand, cold treatment (0 to 9 °C) of sheep reduced digesta mean retention time in the gastrointestinal tract (Barnett et al., 2015) and digestibility of CP (Sano and Terashima, 2001), but did not affect DM digestibility (Sano and Terashima, 2001; Barnett et al., 2015). Nevertheless, cold stress caused an increased rate of passage of digesta and decreased reticulorumen volume, which resulted in reduced digestive efficiency for forage diets in cattle (Young, 1983).

GENE EXPRESSION REGULATION UNDER THERMAL STRESS

Not surprisingly, acute HS and CS induce the expression of several genes, especially heat shock transcription factor 1 (HSF1) and heat shock proteins. Those proteins exert cytoprotective effects to assist in the folding of newly synthesized proteins and repairing and refolding damaged proteins under stress conditions (Kregel, 2002), avoiding cellular apoptosis. When HS persisted, changes in gene expression led to an altered physiological state referred to as "acclimation," a process largely controlled by altered gene expression in response to endocrine signals. The HSF1 is related to thermal tolerance, and there is a high variation among thermal-stressed animals with regard to the expression of this gene (Collier et al., 2008), suggesting that there is opportunity to improve thermotolerance by the manipulation of genes controlling the expression of HSF1.

In goats exposed to HS for 4 weeks, we used microarrays to study gene expression in the whole-blood cells (Salama et al., 2012). We detected many genes whose expression was up- and

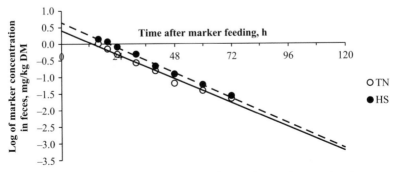

Figure 2.1: Regression of ytterbium concentration in feces of goats kept under thermoneutral (TN) or heat stress (HS) conditions

downregulated by HS. These genes were related to cell proliferation and apoptosis, free radical scavenging, inflammatory response, and glycolysis/gluconeogenesis. In addition, it seemed that the functions of blood immune cells were altered due to changes in their lipid metabolism. This apparent alteration in functionality of immune cells could increase the susceptibility of HS animals to infections, but this assumption needs more investigation.

An RNA sequencing (RNA-seq) analysis of milk somatic cells was carried out (Salama et el., 2015a) in order to investigate whether changes in milk components induced by HS (e.g., lower contents of fat, protein, and lactose) were accompanied by changes in gene expression in the mammary gland. Heat stress downregulated the expression of genes related to milk fat and protein synthesis (Table 2.2). Additionally, the expression of BCL2L1 (anti-apoptotic), and AKT1 (cell survival) decreased by HS, whereas BAX (apoptotic) gene was upregulated by HS (Salama et al., 2015b). Taken together, these results suggest that HS exerts its negative effect on milk production at least in part by inhibiting mammary synthetic capacity as well as by increasing the apoptosis of mammary cells without affecting cell proliferation.

Depending on the intensity and duration of CS, mammalian cells can undergo

Table 2.2 Mammary gene expression as affected by heat stress in goats. The RNA was extracted from milk cells and then the RNA was sequenced by next generation sequencing (RNA-seq). ↑ and ↓ indicate up- and downregulation of genes, respectively by heat stress

Component/gene related	Expression by HS	Gene function
Fat		
ACACA	↓	De novo fatty acid synthesis
FASN	↓	De novo fatty acid synthesis
SCD	↓	Fatty acid desaturation
BTN1A1	↓	Fat globule formation
XDH	↓	Fat globule formation
GLYCAM1	↓	Fat globule membrane protein
Protein and lactose		
CSN1S1	↓	Alpha-S1-casein synthesis
CSN1S2	↓	Alpha-S2-casein synthesis
CSN3	↓	Kappa casein synthesis
LALBA	↓	Lactose synthesis
LTF	↓	Lactotransferrin (innate immunity)
Proteases		
CTSB	↑	Cathepsin B
CTSD	↑	Cathepsin D
CTSZ	↑	Cathepsin Z
PLAU	↑	Urokinase type plasminogen activator
PLAUR	↑	PLAU receptor
UBAP1	↑	Ubiquitin associated protein 1

apoptosis or necrosis. Compared to HS, CS leads to limited effects on gene expression. Heat shock proteins induced by HSF allow cells to adapt to environmental changes and play an important role in stress tolerance and thermal adaptation. Both high and low ambient temperatures induced the expression of HSP70 family genes in blood cells of goats (Banerjee et al., 2014) and buffaloes (Pawar et al., 2014). Sonna et al. (2002) reported that cold exposure of cells (25 to 33°C) activates the expression of specific cold shock proteins such as cold-inducible RNA binding protein 1 (CIRBP) and RNA binding motif protein 3 (RBM3). The CIRBP has a chaperoning role protecting and restoring native RNA conformations during cold stress. Additionally, CIRBP may control the cell cycle as a suppressor of mitosis, and as a molecule involved in maintenance of differentiated states. The RBM3 is a cold-inducible mRNA binding protein that enhances global protein synthesis at both physiological and mild hypothermic temperatures. The DAZ Associated Protein 2 (DAZAP2) has a role in cell signaling and transcription regulation, formation of stress granules during translation arrest, and RNA splicing.

STRATEGIES TO REDUCE THE IMPACT OF THERMAL STRESS IN RUMINANTS: CURRENT SITUATION AND WHAT WE NEED TO LEARN

Generally, there are three main strategies for reducing the negative effects of thermal stress: (1) modification of feeding and the use of feed supplements; (2) environmental modifications and the use of cooling/heating systems; and (3)

genetic development of more thermal-tolerant animals.

Feeding Management: Feeding Schedule and the Use of Feed Supplements

As indicated above, the reduction in feed intake is a typical response to HS and is a threat for one of the five welfare freedoms. Increasing the energy density of the diet could be a solution. This is accomplished by supplementation with concentrates. As reviewed by West (2003), cattle in warm environments selectively decrease the intake of forage to reduce heat production. Increasing the proportion of concentrate in the diet had positive effects on DM intake and on milk energy produced per unit of feed energy (production efficiency). Furthermore, high concentrate rations fed to cattle exposed to severe cold climates can help mitigate cold stress because the extra metabolic energy derived from starch is more advantageous than the extra heat increment derived from fiber. However, supplementation with concentrates in HS conditions should be considered with care as heat-stressed animals are highly prone to ruminal acidosis as indicated above. In addition, feeding high concentrate rations will negatively affect milk fat content in dairy animals. Under HS and CS conditions, feeding high-quality roughage helps maintain feed intake level and productive efficiency of meat and milk. Therefore, caretakers should provide supplemental forage with high quality or move livestock when pasture is limited or of poor quality.

The addition of fat to the diet of ruminants is also a common practice to increase dietary energy density. Feeding fat is associated with reduced metabolic heat production per unit of energy fed and, compared with starch and fiber, it has a much lower heat increment in the rumen (Van Soest, 1982). Gaughan and Mader (2009) reported that feedlot *Bos taurus* beef cattle fed diets supplemented with 1% salt and 5% whole soybeans (as fat source) had elevated tympanic temperature compared with cattle fed diets without supplementations during cold stress (−2.9 to 3.8°C). Furthermore, beef females fed a high-fat diet during late gestation delivered calves with greater resistance to cold stress due to increased brown adipose tissue activity, thermogenesis, and cold resistance (Lammoglia et al., 1999). Similarly, feeding pregnant Angus and Hereford cows on diets supplemented with safflower seed and whole cottonseeds reduced time to standing and increased serum immunoglobulin concentrations of their neonatal calves (Dietz et al., 2003).

HS reduces feed intake (and consequently N intake), and feeding protein above the requirements could be advantageous. However, feeding excess protein is associated with an increase in energy cost related to the deamination of amino acids and the production of urea. Other than the amount of protein fed, quality of protein sources (degradability, and biological value) should be taken into account. If an excess of protein is to be fed to HS animals, this extra amount of protein should come from undegradable sources. Huber et al. (1994) demonstrated that heat-stressed cows fed a highly rumen degradable protein (RDP) diet (65% RDP) had a 6% reduction in DM intake and an 11% decrease in milk yield, when compared with diets of lower degradable protein (59% RDP).

Supplementation with yeast (*Saccharomyces cerevisiae*) during summer increased milk and milk solids yield by 5% and tended to increase blood glucose and niacin levels (Salvati et al., 2015). Furthermore, cows supplemented with yeast had a lower respiratory rate, but had similar rectal temperature compared with non-supplemented animals (Salvati et al., 2015). Supplementation of beef cattle with yeast increased dry matter intake and daily gain, and improved immunity functions (Broadway et al., 2015). The γ-aminobutyric acid (GABA) is an inhibitory neurotransmitter in the central and peripheral nervous systems, and has certain physiological functions such as regulating body temperature and feed intake. Cheng et al. (2014) fed dairy cows with rumen-protected GABA during the summer, and reported a decrease in body temperature coupled with increases in feed intake, milk yield, and milk protein content. Nicotinic acid (niacin), but not nicotinamide, has been known to cause an increase in skin vasodilation, which increases peripheral heat loss (Di Costanzo et al., 1997). Work with dairy cows demonstrated that skin temperatures decreased during periods of mild to severe HS when 12, 24, or 36 g/d of unprotected niacin were supplemented (Di Costanzo et al., 1997). Similarly, supplementation with 12 g/d of rumen-protected niacin resulted in lower vaginal temperatures in lactating dairy cows (Zimbelman et al., 2013). However, Rungruang et al. (2014) tested rumen-protected niacin (4, 8, or 12 g/d) in thermoneutral and HS lactating dairy cows and observed that dietary niacin increased water intake in control and HS cows, without affecting their core body temperature. Furthermore, niacin supplementation did not affect sweating rate, DM intake, or milk yield in either environment. One interesting finding was the decrease in blood niacin concentration under HS conditions, and this decrease was partly corrected by the niacin supplementation. These responses underscore the greater metabolism of niacin under HS conditions and the need for dietary supplementation of niacin during thermal stress.

Environmental Modifications

Environmental modifications are easier in intensive controlled production systems than in the grazing extensive systems in large open areas. In practical conditions, provision of shade with trees or portable structures (animals generally prefer natural shade) is the primary means for reduction of sun radiation. Depending on shade quality, the provision of shade would reduce radiant heat load by 30 to 70% (West, 2003). Armstrong (1994) reviewed the benefits and deficiencies of various types of shade and concluded that shade orientation varies between dry and wet climates. In addition, some basic shade structure recommendations for dairy cow farms can be found in the fact sheet of Fidler and VanDevender (Fidler and VanDevender, University of Arkansas, FSA3040). For beef cattle, the addition of shade structures effectively decreased respiration rates and increased feed intake, growth rate, and final hot carcass weight (Gaughan et al., 2010; Blaine and Nsahlai, 2011).

Under HS conditions, if the air temperature is greater than skin temperature, the air movement will promote the accumulation of heat into the animal rather than the dissipation of heat from the animal (reviewed by Kadzere et al., 2002). Typically, reducing the environmental heat load is based on water sprinklers and fans to either decrease the environmental temperature or to increase the rate of water evaporation from the skin of the animal. Evaporative cooling is affected by wind velocity, relative humidity, and thermal and solar radiation. Other factors that affect the efficacy of evaporative cooling from the skin surface are hair coat density and thickness, hair length and color, and skin color, as previously mentioned. The air velocity remains the most important factor influencing evaporative heat transfer of the skin surface by sweating. Therefore, animals should be maintained in areas with enough air flow to facilitate heat dissipation.

Sprinkling of animals in the morning before getting hot as well as sprinkling or wetting shelter surfaces resulted in lower heat load on beef cattle (Davis et al., 2003). Cooling the surface would appear to provide a heat sink for cattle to dissipate body heat, thus allowing cattle to better adapt to environmental conditions. The use of sprinklers in extensive areas is not easy as animals are freely moving in a large area. However, sprinklers assembling at some points (e.g., water source) is beneficial if the humidity is not high (not more than 60%).

One of the quickest methods of minimizing cold stress is to provide insulation or shelter for the animal. Also, it is essential to keep the animal dry. Using windbreaks (natural or artificial) to protect from wind if animals are prone to getting wet is critical to mitigate the adverse effects of low temperatures. However, windbreaks could reduce air flow in the summer and may increase the need for sprinklers and/or shade to avoid HS (Mader, 2014).

Improving Thermal Tolerance by Genetics

The impact of thermal stress can be relieved by the aforementioned modification of feeding or environment, or by genetic selection of animals less affected by thermal stress. In general, the selection for greater milk or meat production has resulted in animals with greater susceptibility to health disorders and environmental changes (Kadzere et al., 2002). Identification of individual animals that are highly productive and at the same time thermal tolerant would be beneficial if these animals were able to maintain their high productivity and survivability when exposed to adverse climatic conditions. However, the identification of thermal-tolerant animals is not easy under field conditions, but it can be based on measurements of their acute response (e.g., rectal temperature and respiratory rate) to a sudden short exposure to extreme HS or CS. Gaughan et al. (2010) proposed the use of visual panting score (0 to 4.5 scale where 0 = no stress and 4.5 = extreme stress) to evaluate the heat tolerance of 17 beef cattle genotypes. Under high heat load, a greater number of pure bred *Bos taurus* and crosses of *Bos taurus* cattle had a panting score ≥ 2 compared to Brahman cattle, and Brahman-cross cattle. Furthermore, thermal tolerance can be assessed genetically by using animal models. Using the "reaction norm animal model," Menendez-Buxadera et al. (2012) identified those goats that have the same performance throughout a THI range and those with varying performance (tolerant

and non-tolerant to HS). The fact that there is enough genetic variation for the response of milk production traits to HS in goats (Menendez-Buxadera et al., 2012) and cows (Ravagnolo and Misztal, 2000) makes it possible to select for heat tolerance.

Animal variation has been shown to exist for body temperature regulation of beef and dairy cattle during periods of extreme external temperature, with heritability estimates ranging from 0.11 to 0.44 (Dikmen et al., 2012; Howard et al., 2014). The high variation in body temperature heritability might make it feasible to select for more stable body temperatures during summer and winter months to improve resistance to HS and CS. However, those authors demonstrated that selection for body temperature could lead to lower productivity unless methods are used to identify genes affecting rectal temperature that do not adversely affect milk and meat production. It was also possible to improve the reliability of genetic estimates for body temperature via genome-wide association mapping that identifies some key SNPs associated with the regulation of rectal temperature in dairy (Dikmen et al., 2013) and beef (Howard et al., 2014) animals. Similarly, Basiricò et al. (2011) investigated the association between inducible *HSP*70.1 SNPs and heat shock response of peripheral blood mononuclear cells in dairy cows. They identified mutation sites that may be useful as genetic markers for the selection of heat-tolerant animals. Taken together, it seems possible to identify genetic markers (unrelated to milk or meat production) that predict thermotolerance, and selection for these markers could lead to an increase in the resistance to heat and cold without affecting animal productivity.

There are interactions between genetic background (including large effect mutations) and the environment. One mutation is myostatin, which produces an inactive myostatin protein product causing the well-characterized "double muscling" phenotype. An animal with two copies of the inactive myostatin allele yields an extremely lean and heavily muscled carcass with reduced fat cover compared to an animal with one copy (Casas et al., 2004). Myostatin genotype of crossbred Angus and Simmental steers and heifers affected the response to both HS and CS, with 1-copy animals more robust to environmental extremes in comparison with 0- or 2-copy animals (Howard et al., 2013). This information about the genetic × ambient temperature interaction could help in the management of cattle to ensure optimal performance.

Constraints and/or disadvantages of selection for thermal tolerance in ruminants include elevated cost due to the long generation interval. In addition, Ravagnolo and Misztal (2000) demonstrated that good selection not only requires high amounts of data, but also that the data should be accurate (which is not accomplished in many cases, especially in extensive production systems). We should also keep in mind that results of various genetic evaluation studies revealed that, while the thermal tolerance increases, the genetic level for productive traits decreases. Furthermore, selection for heat-tolerant animals could result in animals that are more sensible to cold stress and vice versa, but this genetic antagonism could be circumvented by using marker-assisted selection. Alternative approaches to genetic selection could be crossbreeding with more thermal-tolerant animals. Crossbreeding in dairy cattle has been successful under extensive, but not intensive, management because of lower production levels than in purebreds. Also, while the F1 may be thermal-tolerant, the more complex crossbreds may be less tolerant.

Few specific genes that control heat tolerance have been identified. One of these genes is that one located at the SLICK locus on chromosome 20, and found to control hair length. Originally described in Senepol cattle, the gene was subsequently identified in Carora cattle (Olson et al., 2003) and introduced into different European breeds by crossbreeding. Crossbred Holstein, Angus, and Charolais are as heat tolerant as Brahman (Mariasegaram et al., 2007). Holstein cows inheriting slick hair had lower body temperature and respiration rate than wild-type cows under HS conditions because of greater sweating rates (Dikmen et al., 2008). When submitted to HS the 75:25% Holstein:Carora cows with slick hair had lower rectal temperature (−0.5°C) with numerical greater 305-d milk yield (+285 kg) than Holstein cows (Olson et al., 2003). Recently, Dikmen et al. (2014) confirmed that Holstein cows with slick hair have superior thermoregulatory ability compared with non-slick animals and experience a less drastic depression in milk yield during the summer.

SUMMARY AND CONCLUSIONS

As shown in Figure 2.2, it is important to monitor climate conditions and animals.

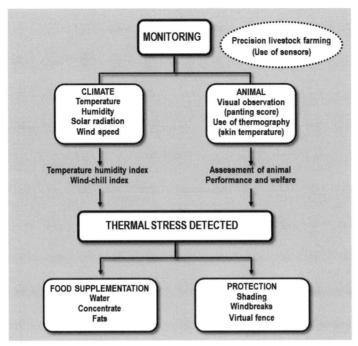

Figure 2.2: Detection of thermal stress and actions that should be taken under extensive production systems. Both climate and animal data should be monitored for the detection of thermal stress situations. Once thermal stress situation is detected, managers should take measures to alleviate the impact of thermal stress on animals.

In summer when heat stress situations are expected, temperature and humidity records are needed for the calculation of a temperature–humidity index with corresponding corrections for solar radiation and wind speed. In cold environments, the wind-chill index indicates the effective temperature. Getting wet and exposure to high velocity of wind in cold environments markedly reduce the effective temperature (real feeling of temperature), which increases the cold stress intensity on animals.

In parallel, animals should be visually monitored, which is not an easy task for animals grazing and freely moving in large open areas. Panting score and measuring skin temperature by thermography could be done at water points. Precision livestock farming (PLF) is a modern approach that can help in the detection of thermal stress situations when animals wear sensors that continuously collect physiological data (e.g., body temperatures or flank movements as indicators of panting). Furthermore, by using the PLF approach, virtual shepherding is possible in which a virtual boundary is defined (as a series of latitude/longitude coordinates), animal location is determined by GPS, and a warning signal (usually electric shock) is emitted if the animal attempts to cross the boundary. By means of the virtual shepherd, managers could direct animals toward the shaded areas of pasture on hot days or wind-protected areas when the weather is cold.

Water should be available as heat-stressed animals depend on evaporative cooling by panting and sweating. In cold environments when water is frozen, water consumption will decline and feed intake will decrease, preventing animals from eating more food needed for extra metabolic heat production, which is required to maintain body temperature. Supplementation with concentrates and fat sources (e.g., whole oilseeds) compensates for the reduction in forage intake on hot days and covers the extra needs for energy on cool days. Concentrates supplementation should be done gradually to avoid subacute ruminal acidosis. One could also consider the use of high energy, non-starch feed stuffs such as distiller grains and soy hulls to avoid large decreases in rumen pH. Additionally, if virtual shepherding is not possible, caretakers should move animals to shaded areas (with trees) on hot days and try to keep animals protected from rain and wind as much as possible on cold days. In the long term, selection for cold- and heat-tolerant animals or crossing with thermal-tolerant breeds could be a good solution. However, for successful selection a large amount of accurate data on animal response to heat and cold stress is needed, which is not readily available in extensive production systems. The application of PLF could offer enough data to better understand the genetic basis of body temperature regulation during extreme temperatures.

Acknowledgements

Part of the results presented in this manuscript come from a research project funded by the Spanish Ministry of Economy and Competitiveness (Plan Nacional I+D+I; Project AGL-2013-44061R).

REFERENCES

Abdoun, K.A., Samara, E.M., Okab, A.B., and Al-Haidary, A.A. (2012) Regional and circadian variations of sweating rate and body surface temperature in camels (Camelus dromedarius). *Animal Science Journal* 83, 556–561.

Acharya, R.M., Gupta, U.D., Sehgal, J.P., and Singh, M. (1995) Coat characteristics of goats in relation to heat tolerance in the hot tropics. *Small Ruminant Research* 18, 245–248.

Allen, J.D., Hall, L.W., Collier, R.J., and Smith, J.F. (2015) Effect of core body temperature, time of day, and climate conditions on behavioral patterns of lactating dairy cows experiencing mild to moderate heat stress. *Journal of Dairy Science* 98, 118–127.

Ames, D.R., and Insley L.W. (1975) Windchill effect for cattle and sheep. *Journal of Animal Science* 40, 161–165.

Armstrong, D.V. (1994) Heat stress interaction with shade and cooling. *Journal of Dairy Science* 77, 2044–2050.

Bandaranayaka, D.D., and Holmes, C.W. (1976) Changes in the composition of milk and rumen contents in cows exposed to a high ambient temperature with controlled feeding. *Tropical Animal Health and Production* 8, 38–46.

Banerjee, D., Upadhyay, R.C., Chaudhary, U.B., Kumar, R., Singh, S., Jagan, A., Mohanarao G., Polley, S., Mukherjee, A., Das, T.K., and De, S. (2014) Seasonal variation in expression pattern of genes under HSP70. *Cell Stress Chaperones* 19, 401–408.

Barnett, M.C., McFarlane, J.R., and Hegarty, R.S. (2015) Low ambient temperature elevates plasma triiodothyronine concentrations while reducing digesta mean retention time and methane yield in sheep. *Journal of Animal Physiology and Animal Nutrition* 99, 483–491.

Basiricò, L., Morera, P., Primi, V., Lacetera, N., Nardone, A., and Bernabucci, U. (2011)

Cellular thermotolerance is associated with heat shock protein 70.1 genetic polymorphisms in Holstein lactating cows. *Cell Stress and Chaperones* 16, 441–448.

Baumgard, L.H., and Rhoads, R.P. (2012) Ruminant production and metabolic responses to heat stress. *Journal of Animal Science* 90, 1855–1865.

Baumgard, L.H., and Rhoads, R.P. (2013) Effects of heat stress on postabsorptive metabolism and energetics. *Annual Review of Animal Bioscience* 1, 311–337.

Bernabucci, U., Bani, P., Ronchi, B., Lacetera, N., and Nardone, A. (1999) Influence of short- and long-term exposure to a hot environment on rumen passage rate and diet digestibility by Friesian heifers. *Journal of Dairy Science* 82, 967–973.

Bernabucci, U., Lacetera, N., Baumgard, L.H., Rhoads, R.P., Ronchi, B., and Nardone, A. (2010) Metabolic and hormonal acclimation to heat stress in domesticated ruminants. *Animal* 4, 1167–1183.

Blaine, K.L., and Nsahlai, I.V. (2011) The effects of shade on performance, carcass classes and behavior of heat-stressed feedlot cattle at the finisher phase. *Tropical Animal Health and Production* 43, 609–615.

Bohmanova, J., Misztal, I., and Cole, J.B. (2007) Temperature-humidity indices as indicators of milk production losses due to heat stress. *Journal of Dairy Science* 90, 1947–1956.

Broadway, P.R., Carroll, J.A., and Burdick Sanchez, N.C. (2015) Live yeast and yeast cell wall supplements enhance immune function and performance in food-producing livestock: A review. *Microorganisms* 3, 417–427.

Brown, D.L., Morrison, S.R., and Bradford, G.E. (1988) Effects of ambient temperature on milk production of Nubian and Alpine goats. *Journal of Dairy Science* 71, 2486–2490.

Bryant, J.R., Matthews, L.R., Davys, J. (2010) Development and application of a thermal stress model. Pages 360–364 in *The Proceedings of the 4th Australasian Dairy Science Symposium*.

Calamari, L., Abeni, F., Calegari, F., and Stefanini, L. (2007) Metabolic conditions of lactating Friesian cows during the hot season in the Po valley. 2. Blood minerals and acid-base chemistry. *International Journal Biometeorology* 52, 97–107.

Carroll, J.A., Burdick, N.C., Chase Jr, C.C., Coleman, S.W., and Spiers, D.E. (2012) Influence of environmental temperature on the physiological, endocrine, and immune responses in livestock exposed to a provocative immune challenge. *Domestic Animal Endocrinology* 43, 146–153.

Casas, E., Bennett, G.L., Smith, T.P.L., and Cundiff, L.V. (2004) Association of myostatin on early calf mortality, growth, and carcass composition traits in crossbred cattle. *Journal of Animal Science* 82, 2913–2918.

Castro-Costa, A., Salama, A.A.K., Moll, X., Aguiló, J., and Caja, G. (2015) Using wireless rumen sensors for evaluating the effects of diet and ambient temperature in non-lactating dairy goats. *Journal of Dairy Science* 98, 4646–4658.

Cheng, J.B., Bu, D.P., Wang, J.Q., Sun, X.Z., Pan, L., Zhou, L.Y., and Liu W. (2014) Effects of rumen-protected γ-aminobutyric acid on performance and nutrient digestibility in heat-stressed dairy cows. *Journal of Dairy Science* 97, 5599–5607.

Collier, R.J., Beede, D.K., Thatcher, W.W., Israel, L.A., and Wilcox, C.J. (1982) Influences of environment and its modification on dairy animal health and production. *Journal of Dairy Science* 65, 2213–2227.

Collier, R.J., Dahl, G.E., and VanBaale, M.J. (2006) Major advances associated with environmental effects on dairy cattle. *Journal of Dairy Science* 89, 1244–1253.

Collier, R.J., Collier, J.L., Rhoads, R.P., and Baumgard, L.H. (2008) Genes involved in the bovine heat stress response. *Journal of Dairy Science* 91, 445–454.

Cook, N.B., Mentink, R.L., Bennett, T.B., and Burgi, K. (2007) The effect of heat stress and lameness on time budgets of lactating dairy cows. *Journal of Dairy Science* 90, 1674–1682.

Coppock, C.E. (1985) Energy nutrition and metabolism of the lactating dairy cow. *Journal of Dairy Science* 68, 3403–3410.

Davis, M.S., Mader, T.L., Holt, S.M., and Parkhurst, A.M. (2003) Strategies to reduce feedlot cattle heat stress: Effects on tympanic temperature. *Journal of Animal Science* 81, 649–661.

Di Costanzo, A., Spain, J.N., and Spiers, D.E. (1997) Supplementation of nicotinic acid for lactating Holstein cows under heat stress conditions. *Journal of Dairy Science* 80, 1200–1206.

Dietz, R. E., Hall, J.B., Whittier, W.D., Elvinger, F., and Eversole, D.E. (2003) Effects of feeding supplemental fat to beef cows on cold tolerance in newborn calves. *Journal of Animal Science* 81, 885–894.

Dikmen, S., and Hansen, P.J. (2009) Is the temperature-humidity index the best indicator of heat stress in lactating dairy cows in a subtropical environment? *Journal of Dairy Science* 92, 109–116.

Dikmen, S., Alava, E., Pontes, E., Fear, J.M., Dikmen, B.Y., Olson, T.A., and Hansen, P.J. (2008) Differences in thermoregulatory ability between slick-haired and wild-type lactating Holstein cows in response to acute heat stress. *Journal of Dairy Science* 91, 3395–3402.

Dikmen, S., Cole, J.B., Null, D.J., and Hansen, P.J. (2012) Heritability of rectal temperature and genetic correlations with production and reproduction traits in dairy cattle. *Journal of Dairy Science* 95, 3401–3405.

Dikmen, S., Cole, J.B., Null, D.J., and Hansen, P.J. (2013) Genome-wide association mapping for identification of Quantitative Trait Loci for rectal temperature during heat stress in Holstein cattle. *Plos One* 8, e69202.

Dikmen, S., Khan, F.A., Huson, H.J., Sonstegard, T.S., Moss, J.I., Dahl, G.E., and Hansen, P.J. (2014) The SLICK hair locus derived from Senepol cattle confers thermotolerance to intensively managed lactating Holstein cows. *Journal of Dairy Science* 97, 5508–5520.

Elkhawad, A.O. (1992). Selective brain cooling in desert animals: the camel (Camelus dromedarius). *Comparative Biochemistry and Physiology* 101, 195–201.

Feng, S., Hub, Q., Huang, W., Ho, C., Li, R., and Tang, Z. (2014) Projected climate regime shift under future global warming from multi-model, multi-scenario CMIP5 simulations. *Global and Planetary Change* 112, 41–52.

Fidler, A.P., and VanDevender, K. Heat stress in dairy cattle. University of Arkansas. Fact sheet FSA3040. Available online: http://www.uaex.edu/publications/PDF/FSA-3040.pdf

Finch, V.A., Dmi'el, R., Boxman, R., Shkolnik, A., and Taylor, C.R. (1980) Why black goats in hot deserts? Effects of coat colour heat exchanges of wild and domestic goats. *Physiological Zoology* 53, 19–25.

Fuquay, J.W. (1981) Heat stress as it affects animal production. *Journal of Animal Science* 52, 164–174.

Gaughan, J.B., and Mader, T.L. (2009) Effects of sodium chloride and fat supplementation on finishing steers exposed to hot and cold conditions. *Journal of Animal Science* 87, 612–621.

Gaughan, J.B., and Mader, T.L. (2014) Body temperature and respiratory dynamics in un-shaded beef cattle. *International Journal of Biometeorology* 58, 1443–1450.

Gaughan, J.B., Mader, T.L., Holt, S.M., Sullivan, M.L., and Hahn, G. L (2010). Assessing the heat tolerance of 17 beef cattle genotypes. *International Journal of Biometeorology* 54, 617–627.

Grandin, T. (1997) Assessment of stress during handling and transport. *Journal of Animal science* 75, 249–257.

Hamzaoui, S. (2014) Heat stress responses in dairy goats and effects of some nutritional strategies for mitigation. Ph.D. dissertation,

Universitat Autònoma de Barcelona, Barcelona, Spain.

Hamzaoui, S., Salama, A.A.K., Albanell, E., Such, X., and Caja, G. (2013) Physiological responses and lactational performances of late lactating dairy goats under heat stress conditions. *Journal of Dairy Science* 96, 6355–6365.

Hansen, P.J. (2004) Physiological and cellular adaptations of zebu cattle to thermal stress. *Animal Reproduction Science* 82–83, 349–360.

Hirayama, T., Katoh, K., and Obara, Y. (2004) Effects of heat exposure on nutrient digestibility, rumen contraction and hormone secretion of goats. *Animal Science Journal* 75, 237–243.

Howard, J.T., Kachman, S.D., Nielsen, M.K., Mader, T.L., and Spangler, M.L. (2013) The effect of myostatin genotype on body temperature during extreme temperature events. *Journal of Animal Science* 91, 3051–3058.

Howard, J.T., Kachman, S.D., Snelling, W.M., Pollak, E.J., Ciobanu, D.C., Kuehn, L.A., and Spangler, M.L. (2014) Beef cattle body temperature during climatic stress: a genome-wide association study. *International Journal of Biometeorology* 58, 1665–1672.

Huber, J.T., Higginbotham, G., Gomez-Alarcon, R.A., Taylor, R.B., Chen, K.H., Chan, S.C., and Wu, Z. (1994) Heat stress, interactions with protein, supplemental fat and fungal cultures. *Journal of Dairy Science* 77, 2080–2090.

IPCC (2013) Climate Change 2013: The Physical Science Basis. Fifth assessment report of the Intergovernmental Panel on Climate Change. Cambridge University Press.

Itoh, F., Obara, Y., Rose, M.T., and Fuse, H. (1998) Heat influences on plasma insulin and glucagon in response to secretogogues in non-lactating dairy cows. *Domestic Animal Endocrinology* 15, 499–510.

Kadzere, C.T., Murphy, M.R., Silanikove, N., and Maltz, E. (2002) Heat stress in lactating dairy cows: a review. *Livestock Production Science* 77, 59–91.

Kregel, K.C. (2002) Heat shock proteins: modifying factors in physiological stress responses and acquired thermotolerance. *Journal of Applied Physiology* 92, 2177–2186.

Kristensen, T., Jensen, C., Østergaard, S., Weisbjerg, M.R., Aaes, O., and Nielsen N.I. (2015) Feeding, production, and efficiency of Holstein-Friesian, Jersey, and mixed-breed lactating dairy cows in commercial Danish herds. *Journal of Dairy Science* 98, 263–274.

Lammoglia, M.A., Bellows, R.A., Grings, E.E., Bergman, J.W., Short, R.E., and MacNeil, M.D. (1999) Effects of feeding beef females supplemental fat during gestation on cold tolerance in newborn calves. *Journal of Animal Science* 77, 824–834.

Lee, D.H.K. (1965) Climatic stress indices for domestic animals. *International Journal of Biometeorology* 9, 29–35.

Lough, D.S., Beede, D.K., and Wilcox, C.J. (1990) Effects of feed intake and thermal stress on mammary blood flow and other physiological measurements in lactating dairy cows. *Journal of Dairy Science* 73, 325–332.

Mader, T.L. (2014) Animal welfare concerns for cattle exposed to adverse environmental conditions. *Journal of Animal Science* 92, 5319–5324.

Mader, T.L., Davis, M.S., and Brown-Brandl, T. (2006) Environmental factors influencing heat stress in feedlot cattle. *Journal of Animal Science* 84, 712–719.

Marai, I.F.M., El-Darawany, A.A., Fadiel, A., and Abdel-Hafez, M.A.M. (2007) Physiological traits as affected by heat stress in sheep–A review. *Small Ruminant Research* 71, 1–12.

Marai, I.F.M., and Haeeb, A.A.M. (2010) Buffalo's biological functions as affected by heat stress - A review. *Livestock Science* 127, 89–109.

Mariasegaram, M., Chase, C.C. Jr., Chaparro, J.X., Olson, T.A., Brenneman, R.A., and Niedz, R.P. (2007) The slick hair coat locus maps to chromosome 20 in Senepol-derived cattle. *Animal Genetics* 38, 54–59.

McDowell, R.E., Moody, E.G., Van Soest, P.J., Lehmann, R.P., and Ford, G.L. (1969) Effect of heat stress on energy and water utilization of lactating cows. *Journal of Dairy Science* 52, 188–194.

Menendez-Buxadera, A., Molina, A., Arrebola, F., Clemente, I., and Serradilla, J.M. (2012) Genetic variation of adaptation to heat stress in two Spanish dairy goat breeds. *Journal of Animal Breeding and Genetics* 129, 306–315.

Mishra, M., Martz, F.A., Stanley, R.W., Johnson, H.D., Campbell, J.R., and Hilderbrand, E. (1970) Effect of diet and ambient temperature-humidity on ruminal pH, oxidation reduction potential, ammonia and lactic acid in lactating cows. *Journal of Animal Science* 30, 1023–1028.

Montanholi, Y.R., Odongo, N.E., Swanson, K.C., Schenkel, F.S., McBride, B.W., and Miller, S.P. (2008) Application of infrared thermography as an indicator of heat and methane production and its use in the study of skin temperature in response to physiological events in dairy cattle (Bos taurus). *Journal of Thermal Biology* 33, 468–475.

NOAA (2009) Cold advisory for newborn livestock (CANL) fact sheet. http://www.wrh.noaa.gov/ggw/canl2/FactSheet.pdf Accessed May 15, 2015.

NRC (2007) *Nutrient requirements of small ruminants: Sheep, goats, cervids, and new world camelids*. National Academy Press, Washington, DC., USA.

O'Brien, M.D., Rhoads, R.P., Sanders, S.R., Duff, G.C., and L.H. Baumgard (2010) Metabolic adaptations to heat stress in growing cattle. *Domestic Animal Endocrinology* 38, 86–94.

Olson, T.A., Lucena, C., Chase Jr. C.C., and Hammond, A.C. (2003) Evidence of a major gene influencing hair length and heat tolerance in Bos taurus cattle. *Journal of Animal Science* 81, 80–90.

Pawar, H.N., Kumar, G.V.P.P.S.R., Narang, R., and Agrawal, R.K. (2014) Heat and cold stress enhances the expression of heat shock protein 70, heat shock transcription factor 1 and cytokines (IL-12, TNF- and GMCSF) in buffaloes. *International Journal of Current Microbiology and Applied Science* 3, 307–317.

Purwanto, B.P., Abo, Y., Sakamoto, R., Furumoto F., and Yamamoto, S. (1990) Diurnal patterns of heat production and heart rate under thermoneutral conditions in Holstein Friesian cows differing in milk production. *Journal of Agricultural Science (Camb.)* 114, 139–142.

Ravagnolo, O., and Misztal, I. (2000) Genetic component of heat stress in dairy cattle, parameter estimation. *Journal of Dairy Science* 83, 2126–2130.

Revell, D.K. (2016) Adaptive responses of rangeland livestock to manage water balance. In: Villalba, J.J. (Ed.) *Animal Welfare in Extensive Production Systems*. 5m Publishing. Sheffield, UK.

Rhoads, R.P., Baumgard, L.H., Suagee, J.K., and Sanders, S.R. (2013) Nutritional interventions to alleviate the negative consequences of heat stress. *Advances in Nutrition* 4, 267–276.

Rungruang, S., Collier, J.L., Rhoads, R.P., Baumgard, L.H., de Veth, M.J., and Collier R.J. (2014) A dose-response evaluation of rumen-protected niacin in thermoneutral or heat-stressed lactating Holstein cows. *Journal of Dairy Science* 97, 5023–5034.

Rutter, S.M. (2014) Smart technologies for detecting animal welfare status and delivering health remedies for rangeland systems. *Scientific and Technical Review of the Office International des Epizooties* 33, 181–187. Available online: http://www.oie.int/doc/ged/D13665.PDF

Salama, A.A.K., Hamzaoui, S., Badaoui, B., Zidi, A., and Caja, G. (2012) Transcriptome analysis of blood in heat-stressed dairy goats. ADSA-ASAS Joint Meeting. Phoenix, Arizona. *Journal of Dairy Science* 95(Suppl. 2), 188.

Salama, A.A.K., Caja, G., Hamzaoui, S., Badaoui, B., Castro-Costa, A., Façanha, D.A.E., Guilhermino, M.M., and Bozzi R. (2014) Different levels of response to heat stress in dairy goats. *Small Ruminant Research* 121, 73–79.

Salama, A.A.K., Badaoui, B., Hamzaoui1, S., and Caja, G. (2015a) RNA-sequencing analysis of milk somatic cells in heat-stressed dairy goats. *Journal of Dairy Science* 98(E-Suppl. 1), 62.

Salama, A.A.K., Duque, M., Shahzad, K., and Loor, J.J. (2015b) Heat stress and amino acid supplementation affected dramatically the expression of genes related to mammary cell activity and number. *Journal of Dairy Science* 98(E-Suppl. 1), 538.

Salvati, G.G.S., Morais Júnior, N.N., Melo, A.C.S., Vilela, R.R., Cardoso, F.F., Aronovich, M., Pereira, R.A.N., and Pereira M.N. (2015) Response of lactating cows to live yeast supplementation during summer. *Journal of Dairy Science* 98, 4062–4073.

Sano, H., and Terashima, Y. (2001) Effects of dietary protein level and cold exposure on tissue responsiveness and sensitivity to insulin in sheep. *Journal of Animal Physiology and Animal Nutrition* 85, 349–355.

Sano, H., Ambo, K., and Tsuda, T. (1985) Blood glucose kinetics in whole body and mammary gland of lactating goats exposed to heat stress. *Journal of Dairy Science* 68, 2557–2564.

Schütz, K.E., Cox, N.R., and Tucker, C.B. (2014) A field study of the behavioral and physiological effects of varying amounts of shade for lactating cows at pasture. *Journal of Dairy Science* 97, 3599–3605.

Sevi, A., and Caroprese, M. (2012) Impact of heat stress on milk production, immunity and udder health in sheep: A critical review. *Small Ruminant Research* 107, 1–7.

Shwartz, G., Rhoads, M.L., VanBaale, M.J., Rhoads, R.P., and Baumgard, L.H. (2009) Effects of a supplemental yeast culture on heat-stressed lactating Holstein cows. *Journal of Dairy Science* 92, 935–942.

Silanikove, N. (2000a) The physiological basis of adaptation in goats to harsh environments. *Small Ruminant Research* 35, 181–193.

Silanikove, N. (2000b) Effects of heat stress on the welfare of extensively managed domestic ruminants. *Livestock Production Science* 67, 1–18.

Silanikove, N., and Koluman, N. (2015) Impact of climate change on the dairy industry in temperate zones: Predications on the overall negative impact and on the positive role of dairy goats in adaptation to earth warming. *Small Ruminant Research* 123, 27–34.

Sipple, P., and Passel, C.F. (1945) Measurements of dry atmospheric cooling in subfreezing temperatures. *Proceedings of the American Philosophical Society* 89, 177–199.

Smith, D.L., Smith, T., Rude, B.J., and Ward, S.H. (2013) Comparison of the effects of heat stress on milk and component yields and somatic cell score in Holstein and Jersey cows. *Journal of Dairy Science* 96, 3028–3033.

Sonna, L.A., Fujita, J., Gaffin, S.L., and Lilly, C.M. (2002) Invited review: Effects of heat and cold stress on mammalian gene expression. *Journal of Applied Physiology* 92, 1725–1742.

Souilem, O., and Varosme, K. (2009) Physiological particularities of dromedary (*Camelus dromedarius*) and experimental implications. *Scandinavian Journal of Laboratory Animal Science* 36, 19–29.

Tajima, K., Nonaka, I., Higuchi, K., Takusari, N., Kurihara, M., Takenaka, A., Mitsumori, M., Kajikawa, H., and Aminov, R.I. (2007)

Influence of high temperature and humidity on rumen bacterial diversity in Holstein heifers. *Anaerobe* 13, 57–64.

Tew, M.A., Battel, G., and Nelson C.A. (2002) Implementation of a new wind chill temperature index by the national weather service. https://www.google.es/url?sa=t&rct=j&q=&esrc=s&source=web&cd=1&cad=rja&uact=8&ved=0CCEQFjAA&url=https%3A%2F%2Fams.confex.com%2Fams%2Fpdfpapers%2F27020.pdf&ei=Bb5UVcTZBay0sAT564A4&usg=AFQjCNHiO-IIzufTH-IqXPGAP-ZGPwS3Cow&bvm=bv.93112503,d.d24 (Accessed May 14, 2015).

Thompson, G.E. (1985) Respiratory system. In: Young, M.K. (Ed.) *Physiology in livestock. Stress.* CRC Press, Inc. Boca Raton, Florida. USA. pp 155–162

University of Missouri (2012) Thermal Aid. http://thermalnet.missouri.edu/Thermal Aid/ (Accessed 24 June 2015).

Van Soest, P.J. (1982) *Nutritional ecology of the ruminant.* Corvallis, OR: O & B Books, Inc.

Verlander, J.W. (1997) Acid-base balance. In: Cunningam J.G. (Ed.) *Textbook of veterinary physiology.* Saunders, Philadephia, Pa. pp 546–554.

Webster, A.J.F. (1974). Heat losses from cattle with particular emphasis on the effects of cold. In: Monteith, J. and Mount, L.E. (Eds.) *Heat losses from animals and man.* Butterworths, London, UK, pp 205–231.

West, J.W. (1999) Nutritional strategies for managing the heat-stressed dairy cow. *Journal of Animal Science* 77(Suppl. 2): 21–35.

West, J.W. (2003) Effects of heat-stress on production in dairy cattle. *Journal of Dairy Science* 86, 2131–2144.

Wheelock, J.B., Rhoads, R.P., VanBaale, M.J., Sanders, S.R., and Baumgard, L.H. (2010) Effects of heat stress on energetic metabolism in lactating Holstein cows. *Journal of Dairy Science* 93, 644–655.

Young, B.A. (1981) Cold stress as it affects animal production. *Journal of Animal Science* 52, 154–163.

Young, B.A. (1983) Ruminant cold stress: effect on production. *Journal of Animal Science* 57, 1601–1607.

Zimbelman, R.B., Collier, R.J., and Bilby, T.R. (2013) Effects of utilizing rumen protected niacin on core body temperature as well as milk production and composition in lactating dairy cows during heat stress. *Animal Feed Science and Technology* 180, 26–33.

chapter three

Adaptive responses of rangeland livestock to manage water balance

Dean K. Revell[1,2]

INTRODUCTION

Water is essential for life and is the largest component of the mammalian body, at 50–80% of body mass depending on age and body fatness. Water is essential for the maintenance of the osmotic pressure of blood, the elimination of waste products, for the production of body fluids including saliva, digestive fluids, and milk, and in thermoregulation (Figure 3.1). For animal production and welfare, it is not so much the amount of water in the body that is critical, but the ability of an animal to replace water that is lost through these processes. Small losses of body water can be more detrimental to an animal than the equivalent percentage losses of any other body component.

Most livestock are more tolerant of body water depletion than humans but, using data from humans as a guide, signs of thirst will be strong when water loss of about 2% of body weight occurs, and a loss of about 5% of the body's water is classified as dehydration, with serious

consequences for metabolic and physiological functions. Losses of 15% and 20% can lead to coma and death. Hence, the avoidance of dehydration is critical for survival, welfare, and productivity.

The proportion of body water that is used each day is termed water turnover. To avoid dehydration, water intake should equal water turnover. Animals can experience short-term deviations in water balance (intake minus loss) without major consequences but a water deficit cannot be maintained in the longer term without ill effects. Water turnover varies between animal species. Animals suited to arid environments have a lower water turnover, with differences between ruminant species being up to two- to three-fold (King, 1979), although the differences between cattle and sheep are not always large (Macfarlane and Howard, 1972). Importantly, water turnover can also vary for an individual animal. The adaptability of an animal offers a potential tool in production systems to help livestock adapt to fluctuations in water availability and their thermal environment.

[1] Revell Science, Duncraig, Western Australia
[2] School of Animal Biology, The University of Western Australia, Nedlands, Western Australia

This chapter will consider four areas, focussing on adaptive responses of livestock and the implications for management to improve welfare and production: 1) A brief overview of the main factors affecting water turnover; 2) A summary of how an animal can conserve body water; 3) How an animal can adapt to environments where the availability of water is limited or their requirements for water are high due to an increase in exposure to radiant heat or high ambient temperatures; and 4) Implications for management of rangeland livestock.

The main factors affecting water intake are ambient temperature and radiant heat, which impact the heat load on an animal, the spatial and temporal availability of water, feed intake, forage type, and water quality (Figure 3.1). The main adaptive responses to an increase in ambient temperature are to increase evaporative cooling and changing habitat selection to reduce the heat load. In response to an increase in water turnover, ruminant livestock are able to concentrate their urine to avoid dehydration (Figure 3.1). New options for managing livestock that may be faced with challenges to their capacity to meet water requirements include behaviour-based strategies and early-life programming.

THE MAIN FACTORS AFFECTING WATER BALANCE AND THE REQUIREMENT TO DRINK WATER

Components of Water Balance

The water drunk by an animal typically comprises about 70–90% of its total water intake, with the remainder arising from metabolic water and from food. Metabolic water is that produced by oxidation of feedstuffs, and is 0.4–1.1 mL/g of tissue oxidised. Lush vegetation can contain up to 90% water, which reduces the need for animals to drink water; in some cases no water needs to be drunk as sufficient is obtained from food. Water loss from the body under thermoneutral conditions is dominated by water excreted in faeces and urine; about one half of total water loss occurs in faeces and about one third in urine with non-lactating animals. Ruminant livestock have only a very limited capacity to reduce water loss by excreting faeces with a higher-than-normal dry matter content. Cattle faeces usually contain 75–85% water and sheep faeces about 67%. The only way they can substantially reduce water loss via faeces is by eating less food and excreting less faecal material. Water lost in the urine, on the other hand, can be regulated to a significant degree. The renin-angiotensin system (RAS) provides multiple control points for water balance, principally by regulating renal sodium and water absorption, and thereby altering the osmolarity of urine. Water lost via sweat is normally about 15% of water turnover, but it can be considerably higher, up to 80%, under hot conditions. The lactating animal has a particularly challenging situation, as it is excreting a product, milk, containing about 85% water, and its high metabolic rate is producing heat that needs to be dissipated (unless in a cold environment). The combined effects of water in milk and water lost in thermoregulation mean that every litre of milk requires an extra 3 litres of water to be consumed (Barrett and Larkin, 1974).

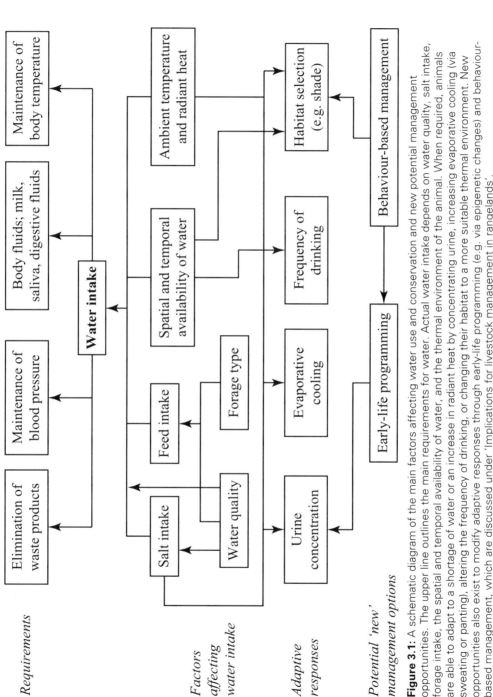

Figure 3.1: A schematic diagram of the main factors affecting water use and conservation and new potential management opportunities. The upper line outlines the main requirements for water. Actual water intake depends on water quality, salt intake, forage intake, the spatial and temporal availability of water, and the thermal environment of the animal. When required, animals are able to adapt to a shortage of water or an increase in radiant heat by concentrating urine, increasing evaporative cooling (via sweating or panting), altering the frequency of drinking, or changing their habitat to a more suitable thermal environment. New opportunities also exist to modify adaptive responses through early-life programming (e.g. via epigenetic changes) and behaviour-based management, which are discussed under 'Implications for livestock management in rangelands'.

Ambient Temperature

The two main factors affecting water turnover are ambient temperature and the level of feed intake. These two factors are related to each other, as higher temperatures lead to a reduction in feed intake when evaporative cooling mechanisms are insufficient to manage body temperature. This means that the effect of high temperature on water balance is partially offset by a lower heat increment associated with a reduction in feed intake. Water consumption increases as ambient temperature increases (Figure 3.2), although in any one study, ambient temperature accounts for only 50–84% of the variation in water intake, so other factors such as feed intake, type of forage, access to shade, coat thickness, and genotype are also important, especially at lower ambient temperatures. At high ambient temperatures, there is a closer relationship between temperature and water intake because of the positive relationship between cooling mechanisms and water requirements. It is generally accepted that sheep have a lower water turnover on a body weight basis than cattle, but the differences are not necessarily large (Figure 3.3).

Feed Intake and Forage Type

Water consumption of cattle, sheep, and goats is positively correlated to feed dry matter (DM) intake. At temperatures less than 25°C, water requirements range from about 4 to 8 L water/kg DM intake for cattle, and 2 to 6 L water/kg DM intake for sheep and goats (reviewed by CSIRO, 2007). At higher temperatures, water intake per unit of feed DM intake can double, with values of 16 L/kg DM intake reported with *Bos taurus* cattle at 38°C.

Forage type effects water requirements in two main ways. The first is the DM content of the forage, which can vary considerably between seasons. In rangelands where succulent plants are available

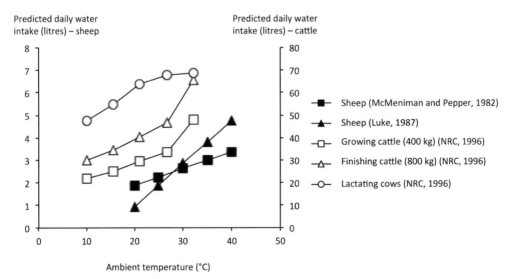

Figure 3.2: Relationships between ambient temperature and water intake for sheep and cattle. Data for sheep are presented on the left-hand axis, and for cattle on the right-hand axis

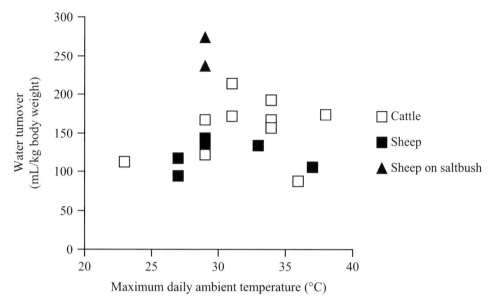

Figure 3.3: Water turnover in sheep and cattle from various studies. Data taken from papers reviewed by CSIRO (2007)

(e.g., cacti), then livestock can acquire a considerable portion of their daily water intake from these. The second way that forage type can affect water requirements is via the salt content of the edible material. For example, sheep grazing the halophytic shrub, *Atriplex nummularia* (old man saltbush), which contains high concentrations of salt (up to 30% NaCl on a DM basis; Ben Salem et al., 2010) have a water turnover approximately double that of sheep on more typical low-salt forages; 173 versus 111 mL/kg body weight for Merno wethers and 350 versus 172 mL/kg body weight for Border Leicesters (Figure 3.3; Macfarlane et al., 1967).

Evaporative Losses

Evaporative cooling is an efficient mechanism for losing body heat under conditions of low relative humidity. In cattle, sweating contributes two thirds of the evaporative loss, with panting contributing one third. The reverse occurs with sheep and goats, with 40% of evaporative heat loss attributable to sweating and 60% to panting. For every litre of water that vaporises, via panting or sweating, 2.4 MJ of heat are lost. If animals are exposed to conditions where they accumulate a "heat load" (Gaughan et al., 2008), they will have a high reliance on evaporative cooling that in turn will increase water turnover and their requirement to drink. Water loss from evaporation can account for up to 80% of the water used by a ruminant in the tropics (King, 1983). A linear relationship has been found between ambient temperature or solar radiation and body water turnover in cattle (King et al., unpublished data, cited by King, 1983), but only up to a point (reached at an average ambient temperature of 29°C or solar radiation of 505 Wm^{-2}) that is limited by an animal's capacity for water turnover. That is, if water loss via evaporation was to continue unabated at high temperatures, it could lead to severe dehydration even with drinking daily (King, 1983).

Water Availability and Quality

The spatial distribution of water can affect grazing patterns and forage utilisation (Ganskopp, 2001) and therefore the composition and abundance of vegetation in the landscape. However, the reverse can also occur; that is, feed quality, availability, and spatial distribution can affect the frequency that animals visit watering points (Squires and Wilson, 1971; Squires, 1976). Therefore, there is a dynamic relationship between feed resources in a rangeland landscape and water utilisation. The temporal and spatial patterns of water usage are made more complex by behavioural patterns of livestock. Individuals and groups form grazing and drinking patterns based on past experiences and modified by current circumstances. This means that strict rules on how the spatial distribution of water points "dictates" drinking frequency need to be treated with some caution; a point discussed in more detail below (see 'Implications for livestock management in rangelands').

In semi-arid and arid rangelands, the frequency of drinking water can actually decline when the pasture dries off, a time when the requirements for drinking water are probably increasing due to higher temperatures. The need to graze further from water points increases the distance animals must walk each day, and reduces the frequency of drinking. When animals must drink less frequently than once a day, feed intake typically declines and body weight gain

can be compromised (see review by Holechek, 1997). Impaired production is partly a consequence of more time spent walking, but also a change in diet composition, as it has been shown sheep on a salt desert range that watered daily ate a greater variety of plants, ate more dry material, and grazed more quietly than counterparts that drank less frequently (Hutchings, 1946, cited by Holechek, 1997). There are examples of dairy cows adapting to a reduction in water availability from *ad libitum* to once or twice a day. (King and Stockdale, 1981). The extent of adaptation under rangeland conditions, where long distances must sometimes be covered to access forage and water, is less clear. Nevertheless, there are reports of rangeland livestock drinking only once every three days, indicating a capacity to adapt to less frequent access to water. One might expect that under most circumstances, productivity will drop if watering frequency declines to much less than daily.

Livestock welfare and productivity, and patterns of vegetation utilisation, are therefore influenced by the spatial and temporal availability of water, but additionally these factors are influenced by water quality. Groundwater can often contain dissolved salts, with the predominant cations usually being sodium and magnesium and the predominant anions bicarbonate, chloride, and sulphate. Guidelines have been established to indicate the suitability of saline water for drinking (e.g., summarised by CSIRO, 2007). Water with total soluble salts of < 5,000 mg/L is considered suitable for sheep and cattle of all ages. Between 5,000–10,000 mg/L, care is required with young or lactating livestock, whilst above 10,000 mg/L, caution is required with unaccustomed cattle but it is suitable for mature sheep given their higher capacity to concentrate urine.

HOW ANIMALS CAN ADAPT TO AVOID DEHYDRATION

Behavioural Responses

Animal behaviour is a powerful force; it is based on experiences and continuously modified by feedback mechanisms, allowing animals to juggle the challenge of preparing for expected conditions based on what has occurred in the past, whilst remaining flexible to adapt to new circumstances. Behavioural change can include physiological and physical responses. The following description on animal responses to an increase in heat load is based on the review by Silanikove (2000), who described a set of stages through which animals progress. As the temperature increases above a range providing "thermal comfort," animals respond physiologically through vasodilation to increase heat loss by radiation, and by sweating and panting to increase evaporative cooling. This phase is described as the "innocuous stage" as homeothermy is attained without difficulty, and fitness and productivity of the animal is not normally compromised. The next phase is the "aversive stage," when evaporative cooling mechanisms are intensified, and water consumption per unit of DM intake accelerates rapidly. Animals cope with maintaining body temperature during this period, but with some difficulty. As the heat load increases further, animals enter the "noxious stage," in which coping mechanisms are inadequate and fitness is impaired.

If the increase in heat load is gradual, animals adapt by reducing feed intake, altering endocrine profiles to reduce basal metabolism, or by moving to a more suitable environment, if that is possible. Animals can lose long-wave radiation from their body where the surrounding environment is cooler than their body temperature, which is more likely to occur in areas with green vegetation and shade. Although green vegetation reflects less of the incoming radiation than bare sand (25 versus 30–40%; Barry and Chorley, 1971, cited by King, 1983), it does not heat up as much due to the cooling effect of transpiration. Where high ambient temperatures occur for a lengthy period of time (e.g., during a hot dry season), areas of green vegetation or moist ground may not be available, and access to shade can be restricted in some rangeland areas depending on the type of vegetation, which will compromise the capacity of livestock to tolerate high temperatures.

Beyond the stages where animals can cope with increasing heat loads, they enter an "extreme stage" where heavy panting and maximal sweating are employed in a final attempt to halt the rise in body temperature. As a consequence of accelerated biochemical processes and the energetic cost of panting, there is an inevitable increase in heat production, so animals enter a vicious cycle leading to heat stroke and death unless the heat load is rapidly removed.

Conserving Body Water

When animals are required to conserve water, they have a capacity to concentrate their urine, which differs between animal species. Desert-adapted animals are able to excrete more concentrated urine than less-adapted animals. Individual animals can also adapt by regulating the concentration of their urine through control mechanisms of the renin-angiotensin system. The retention of sodium from the kidneys is controlled by aldosterone, while arginine vasopressin (AVP) is primarily involved in the regulation of water re-absorption (Figure 3.4). In the distal nephrons of the kidney, fluid has a lower osmolality than plasma (which is about 300 mOsm) because of removal of NaCl in earlier segments of the loops of Henle and active transport in the last sections of tubule (Reece, 2004). Tubular fluid osmolality can be reduced to 150 mOsm. Despite the low osmolality, water can be returned from the tubules to extracellular fluid and thus rehydrate body fluids. The permeability of the tubules to water is controlled by the hormone AVP, and the release of AVP is very sensitive to changes in plasma osmolality of just a few percent, either up or down (Reece, 2004), indicating the importance of maintaining the osmolality of fluids in the body for normal function. Through the actions of AVP and aldosterone, both of which are under the control of renin, animals are able to increase the osmolality of urine to conserve water. With water deprivation, cattle can increase urine osmolality up to 1100 mOsm and sheep up to 3000 mOsm; that is, to about 4 and 10 times the osmolarity of plasma.

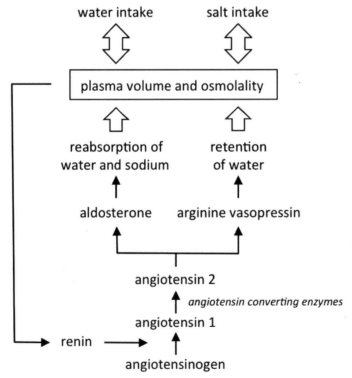

Figure 3.4: The renin-angiotensin system (RAS) with multiple points of regulation that is responsible for the maintenance of water and salt balance (adapted from Blache and Revell, 2016)

Relatively new data has shown that the RAS can be "programmed" to be more or less responsive to changes in salt or water supply. In a series of experiments in Australia (Digby et al., 2008, 2010 a, b; Chadwick et al., 2009 a, b, c; Tay et al., 2012), pregnant ewes were fed a high-salt diet during pregnancy and the physiological consequences on the weaned offspring were monitored. Here, the offspring are subsequently referred to as "high-salt offspring." These studies, together with other reports suggest two types of fetal programming (Digby et al., 2010a). The first is programming of the thirst threshold. When weaned "high-salt offspring" received an oral dose of salt, the increase in water consumption over the following two hours was less than in control animals (i.e., those born from ewes not exposed to salt during pregnancy) (Digby et al., 2010b). In another study, the consumption of water by "high-salt offspring" (at 8 months of age, post-weaning) was about one third less, and water consumption per unit of feed intake was up to 20% less, than control offspring.

The second type of programming appears to be associated with the sensitivity of the RAS. Following an oral salt challenge, the change in aldosterone concentration was blunted in weaned "high-salt offspring," and basal levels of aldosterone were higher than control offspring (Digby et al., 2010b), suggesting "high-salt offspring" are able to retain more salt. This was supported by a separate study, in which "high-salt offspring" had a smaller response of renin to an oral salt dose and tended to retain more salt than normal sheep (Chadwick et al., 2009b). Furthermore, the plasma concentration of AVP was lower in "high salt offspring", suggesting that the sensitivity of the kidney to changes in osmotic pressure was reduced in offspring born to ewes fed high salt during pregnancy (Digby et al., 2010a). This could have very important consequences for the resilience of animals to circumstances that challenge their water and salt balance.

The transfer of adaptive responses to the next generation can occur through epigenetic changes that modify the expression of genes via DNA methylation and histone modification. In the case of "high-salt offspring," DNA methylation in kidney and lung tissue was 20–30% higher compared to control lambs (Tay et al., 2011). The RNA expression of different components of the RAS can be affected by a high-salt diet during gestation (Tay et al., 2011, Mao et al., 2013). The intensity of the exposure to high salt during gestation and the timing of exposure may determine which component of the RAS is affected.

The skin may have a surprising role in salt and water balance (Rabelink and Rotmans, 2009), besides possessing sweat glands that are required for evaporative cooling from the body surface. Substantial sodium retention might occur in skin, without the expected fluid retention and weight gain (Heer et al., 2000). The skin, therefore, can act as an osmotically inactive sodium reservoir, where sodium can be stored when its supply is in excess, and from where sodium can be released during periods of sodium shortage. The plants on offer in many rangeland environments are low in sodium, so using skin as a back-up system to store sodium whenever it can, may be a very useful strategy for health and survival. The non-osmotic storage of sodium in skin allows dissociation of sodium and water handling (Rabelink and Rotmans, 2009) and, indeed, long-term salt restriction can lead to a loss of skin sodium without the loss of water (shown in rats; Schafflhuber et al., 2007), which would assist adaptation to dehydration. If livestock in rangeland environments are exposed to conditions with fluctuations in salt (sodium) and water availability, then mechanisms that regulate salt movement in and out of skin may be very important in understanding adaptive responses.

IMPLICATIONS FOR LIVESTOCK MANAGEMENT IN RANGELANDS

Maintaining Animals in their Zone of Thermal Wellbeing

Good management of rangeland animals will aim to limit the occasions where animals are exposed to the "noxious stage," and more time in the "innocuous" stage of the zone of thermoneutrality (Silanikove, 2000). There are two parts to achieving this. The first is to provide, as much as is possible in rangeland environments, locations where animals can avoid accumulating a heat load or can dissipate a heat load. Principally this relies on providing areas where animals can wet their skin, maintaining areas of green vegetation for as long as possible into dry seasons (e.g., by resting areas during the preceding wet season), and making use of trees that provide sufficient amounts of shade (Rosselle et al., 2013). The second aspect is to manage the livestock to establish experiences of a wide range of areas so they are familiar with locations where thermal comfort can more easily be attained. The area that an animal is reared has a strong influence on that animal's selection of

the same area later in life, although when water ability becomes scarce, the affinity for the "home range" weakens (Howery et al., 1997). This means that when seasonal conditions become difficult (e.g., hot and dry), livestock may be forced to graze further afield in areas unfamiliar to them, making decisions on diet and habitat selection harder for them because they have few, if any, experiences of the new area. Therefore, exposure to multiple areas when animals are young, ideally with their mothers if possible, and when not under nutritional or thermal stress, can predispose the animals to make more informed decisions about diet and habitat when the needs arise.

Across a range of production areas there are anecdotes of animals not using shade and shelter even though it is available. This seems perplexing until one realises that the selection of a habitat is, in part, a function of prior experiences. A sudden change in conditions, such as heat, may require animals to make large changes in habitat selection to avoid accumulating a heat load or to access drinking water more frequently but, without appropriate experience, livestock may not be able to make appropriate decisions. Using peer animals as "trainers" or "guides" can be a practical and useful way to modify grazing behaviours of livestock that are new to an area (Howery et al., 1997; Thomas et al., 2010). Opportunities to influence habitat selection are discussed in more detail under "Influencing grazing behaviours through management."

Genetic Regulation

Selecting animals for suitability to the environment has been a key goal since animals were domesticated. The suitability of cattle to tropical or subtropical areas was reviewed over 65 years ago (Bonsma, 1949), where factors such as coat and skin colour and coat type (e.g., smooth or furry coats) and consequences for production and reproduction were discussed. *Bos indicus* cattle have a higher heat tolerance than *B. taurus*, due largely to their greater ability to sweat and lose heat by evaporative cooling (Vercoe et al., 1972). British breeds tend to drink more than Brahmans (Colditz and Kellaway, 1972). Even within breeds, difference in heat tolerance can be quantified (Gaughan et al., 2010). Genetic differences have also been reported for sheep breeds; for example, Merino wethers had 35–44% lower daily water turnovers than Border Leicester wethers (Macfarlane et al., 1967), which may be partly a consequence of lower feed intake in the Merino wethers. Interestingly, at least in high-production dairy cows, selection for production is associated with a reduced tolerance to heat but, because the association is weak, selection for both traits (production and heat tolerance) is considered feasible (Ravagnolo and Misztal, 2000).

WHAT ELSE DO WE NEED TO KNOW ABOUT LIVESTOCK WELFARE AND WATER BALANCE?

Epigenetic Responses

Animals born and reared in a particular environment often perform better than those that are naive (Provenza, 2008). If broilers are exposed to high environmental temperatures during the neonatal period, they are better able to resist heat stress (May, 1995). On the contrary, heat stress during gestation in pigs compromised future thermoregulatory responses to high ambient temperatures (Johnson et al., 2013). The possibility of thermal preconditioning has not yet been tested in sheep or cattle.

In terms of the physiological control of water balance, there is evidence from studying the consequences of high-salt diets during pregnancy in sheep (Digby et al., 2008, 2010 a, b; Chadwick et al., 2009 a, b, c; Tay et al., 2012) and cattle (Mohamed and Phillips, 2003) that events during pregnancy have postnatal consequences on how an animal regulates salt and water. Restricting water access during pregnancy in an attempt to improve water conservation mechanisms in the offspring is not recommended, for multiple reasons including the welfare of the pregnant females and the physiological importance of water in maintaining a successful pregnancy. However, an approach that may be worth exploring is increasing the dietary salt load during discrete windows of time during pregnancy.

Salt is a relatively cheap supplement, and in most cases, rangeland animals readily consume salt-based supplements unless salt is already freely available in the natural environment. In sheep, salt at 15% of the diet fed throughout pregnancy did not have any detrimental effects on reproductive performance (Digby et al., 2008). Livestock won't select such a high level of dietary salt if they can avoid it, but it is possible that salt intake during pregnancy at more modest concentrations of the diet could still program the regulatory systems for salt and water balance in the offspring. The actual required intake of salt, the "ideal" timing of an increased salt intake during pregnancy, and responses in

the offspring under production conditions requires more research to test the
potential of this suggestion. Responses
in offspring that should be monitored
include drinking patterns (including thirst
thresholds), their capacity to retain water
during periods of water restriction, and
responses in water turnover during high
temperatures. In addition, rumen buffering capacity and the retention of elements
in tissues and fluids should be assessed,
as there is evidence that prolonged supplementation with NaCl can eventually
lead to disturbances in element accumulation (Phillips et al., 2015). It is also critically important to note that high levels of
salt in drinking water should be avoided
during pregnancy, as animals are much
less able to cope with a salt load adminis-
tered in drinking water than in food, and
the consumption of saline water during
pregnancy can have negative outcomes
(Potter and McIntosh, 1974).

Using early-life programming as a management tool is a relatively new concept,
but offers a faster approach than genetic
selection to respond to expected circumstances in the short to medium term.
Early-life programming will not replace
genetic selection, but it offers an additional management option.

Influencing Grazing Behaviours through Management

The temporal and spatial patterns of
water and vegetation utilisation are an
expression of animal behaviours, and
should not be seen as fixed or indefi-
nitely entrenched. This is discussed
in detail in this book in Chapter 4.
Importantly, in terms of livestock management, animal behaviour can be modified, either through learning from other

animals, chance events, or by management. Domestic herbivores use sight
and sound cues, learned diet preferences and aversions, a predisposition
to novel stimuli, and spatial memory to
inform their decisions on habitat and diet
selection (Launchbaugh and Howery,
2005). Launchbaugh and Howery (2005)
concluded "that the deliberate and careful modification of animal attributes and
habitat characteristics could yield options
for rangeland management."

Based on principles that explain
animal behaviours, practical interventions
are being applied in the rangelands of
Australia in a project exploring the potential to positively influence grazing distribution of livestock (Revell et al., 2015). The
principles and procedures are referred
to as "Rangelands Self Herding" or
"Rangelands Self Shepherding" (www.
selfherding.com), and their use by managers allows livestock to develop new
animal experiences that consequently
shape grazing range and utilisation of
water points. With the realisation that
grazing and drinking patterns can be modified, guidelines on the spatial distribution
of water points should be taken as just
that: guidelines, rather than firm rules.

Grazing management that allows
animals to learn about the full array of
nutritional and thermal environments
in their range will help animals adapt to
changing circumstances. This is evident
from a nutritional point of view from the
review of Fynn (2012), who concluded
that the performance of livestock (and
wildlife) during droughts is positively
related to plant diversity, but only if the
animals are familiar with the functional
characteristics of the various components
of the feed base. This conclusion could
perhaps be expanded to include familiarity

with the thermal environments and water availability of different locations they can access. Animals with experience will behave differently than naive animals.

CONCLUSION

The consumption of an adequate amount of water is essential to livestock welfare and productivity. As the heat load on animals increases, the demand for water increases, principally because of the need for water in evaporative cooling mechanisms. Water consumption patterns are not fixed, but depend on the type of feed offered (especially its dry matter content and salt content), feed intake, ambient conditions, and water availability. Productivity and well-being of animals may be compromised if water access is restricted to less than once a day, but this is a guide only as adaptive mechanisms exist that allow animals to increase their tolerance to high temperatures or to conserve body water when an increase in demand for water turnover cannot be matched by an increase in water intake. These adaptations include a capacity to increase the concentration of urine and to modify patterns of food and water consumption. Good management practice will aim to ensure that good quality water is accessible and that animals are provided with an environment where they can choose a local habitat to avoid accumulating a heat load. New approaches to improve the capacity of livestock to cope with challenging conditions include early-life programming to modify the regulation of water balance through the renin-angiotensin system and behaviour-based management to increase the breadth of positive experiences that livestock accumulate in diverse rangeland landscapes.

REFERENCES

Barrett, M.A., and Larkin, P.J. (1974) *Milk and beef production in the tropics.* Oxford University Press, Oxford, UK.

Barry, R.G., and Chorley, R.J. (1971) *Atmosphere, weather and climate*, 9th edn. Routledge, Oxon, UK.

Ben Salem, H., Norman, H.C., Nefzaoui, A., Mayberry, D.E., Pearce, K.L. and Revell, D.K. (2010) Potential use of oldman saltbush (*Atriplex nummularia* Lindl.) in sheep and goat feeding. *Small Ruminant Research* 91, 13-28.

Blache, D., and Revell, D.K. (2016) Short and long term consequences of high salt loads in breeding ruminants. In: V.R. Squires and H.M. El Shaer (Eds.) *Halophytic and salt-tolerant feedstuffs. Impacts on nutrition, physiology and reproduction of livestock.* CRC Press, Boca Raton, Florida, USA. pp. 316–335.

Bonsma, J.C. (1949) Breeding cattle for increased adaptability to tropical and subtropical environments. *The Journal of Agricultural Science* 39, 204–221.

Chadwick, M.A., Vercoe, P.E., Williams, I.H., and Revell, D.K. (2009a) Feeding pregnant ewes a high-salt diet or saltbush suppresses their offspring's postnatal renin activity. *Animal* 3, 972–979.

Chadwick, M.A., Vercoe, P.E., Williams, I.H., and Revell, D.K. (2009b) Dietary exposure of pregnant ewes to salt dictates how their offspring respond to salt. *Physiology and Behaviour* 97, 437–445.

Chadwick, M.A., Vercoe, P.E., Williams, I.H., and Revell, D.K. (2009c) Programming sheep production on saltbush: adaptation of offspring from ewes that consumed high amounts of salt during pregnancy and early lactation. *Animal Production Science* 49, 311–317.

Colditz, P.J., and Kellaway, R.C. (1972) The effect of diet and heat stress on feed intake, growth and nitrogen metabolism in Friesian, F1 Brahman x Friesian, and Brahman heifers. *Australian Journal of Agricultural Research* 23, 717–725.

CSIRO (2007) Nutrient requirements of domesticated ruminants. CSIRO Publishing, Collingwood, Victoria, Australia.

Digby, S.N., Blache, D., Masters, D.G., and Revell D.K. (2010a) Responses to saline drinking water in offspring born to ewes fed high salt during pregnancy *Small Ruminant Research* 91, 87–92.

Digby, S.N., Masters, D.G., Blache, D., Hynd P.I., and Revell D.K. (2010b) Offspring born to ewes fed high salt during pregnancy have altered responses to oral salt loads. *Animal* 4, 81–88.

Digby, S.N., Masters, D.G., Blache, D., Blackberry, M.A., Hynd, P.I. and Revell, D.K. (2008). Reproductive capacity of Merino ewes fed a high-salt diet. *Animal* 2, 1353-1360.

Fynn, R.W.S. (2012) Functional resource heterogeneity increases livestock and rangeland productivity. *Rangeland Ecology and Management* 65, 319–329.

Ganskopp, D. (2001) Manipulating cattle distribution with salt and water in large arid-land pastures: A GPS/GIS assessment. *Applied Animal Behavioural Science* 73, 251–262.

Gaughan, J.B., Mader, T.L., Holt, S.M., and Lisle, A. (2008) A new heat load index for feedlot cattle. *Journal of Animal Science* 86, 226–234.

Gaughan, J.B., Mader, T.L., Holt, S.M., Sullivan, M.L., and Hahn, G.L. (2010) Assessing the heat tolerance of 17 beef cattle genotypes. *International Journal of Biometeorology* 54, 617–627.

Heer, M., Baisch, F., Gerzer, R., Drummer, C., and Kropp, J. (2000) High dietary sodium chloride consumption may not induce body fluid retention in humans. *American Journal of Physiology – Renal Physiology.* 278, F585–F595.

Holechek, J.L. (1997) The effects of rangeland water developments on livestock production and distribution. *Proceedings of a Symposium on Environmental, Economic, and Legal Issues Related to Rangeland Water Development.* ASU College of Law, Tempe, Arizona, USA, 13–15 November 1997.

Howery, L.D., Provenza, F.D. Banner, R.E., and Scott, C.B. (1997) Social and environmental factors influence cattle distribution on rangeland. *Applied Animal Behaviour Science* 55, 231–234.

Hutchings, S.S. (1946) Drive the water to the sheep. *National Wool Grower* 36: 10–11.

Johnson, J.S., Boddicker, R.L., Sanz-Fernandez, M.V., Ross, J.W., Selsby, J.T., Lucy, M.C., Safranski, T.J., Rhoads, R.P., and Baumgard, L.H. (2013) Effects of mammalian *in utero* heat stress on adolescent body temperature. *International Journal of Hyperthermia* 29, 696–702.

King, J.M. (1979) Game domestication for animal production in Kenya: field studies of the body-water turnover of game and livestock. *The Journal of Agricultural Science (Cambridge)* 93, 71–79.

King J.M., (1983) Livestock Water Needs in Pastoral Africa. Research Report No 7. ILCA, Addis Ababa.

King, J.M., and Stockdale, C.R. (1981) *Australian Journal of Experimental Agriculture and Animal Husbandry* 21, 167–171.

Launchbaugh, K.L., and Howery, L.D. (2005) Understanding landscape use patterns of livestock as a consequence of foraging behaviour. *Rangeland Ecology and Management* 56, 99–108.

Luke, G.J. (1987) Resource Management Technical Note 60, pp. 1–22. Department of Agriculture: Western Australia.

Macfarlane, W.V., and Howard, B. (1972) Comparative water and energy economy of wild and domestic mammals. *Symposium of the Zoological Society of London* 31, 261–296

Macfarlane, W.V., Howard, B.H., and Siebert, B.D. (1967) Water metabolism of Merino and Border Leicester sheep grazing saltbush. *Australian Journal of Agricultural Research* 18: 947–959.

Mao, C., Liu, R., Bo, L., Chen, N., Li, S., Xia, S., Chen, J., Li, D., Zhang, L., and Xu, Z. (2013) High-salt diets during pregnancy affected fetal and offspring renal renin–angiotensin system. *Journal of Endocrinology* 218, 61–73.

May, J.D. (1995) Ability of broilers to resist heat following neonatal exposure to high environmental temperature *Poultry Science* 74, 1905–1907.

McMeniman, N.P., and Pepper, P.M. (1982) *Proceedings of the Australian Society of Animal Production* 14, 443–446.

Mohamed, M.O., and Phillips, C.J.C. (2003) The effect of increasing the salt intake of pregnant dairy cows on the salt appetite and growth of their calves. *Animal Science* 77, 181–185.

NRC (1996) National Research Council Nutrient Requirements of Beef Cattle, 7th revised edn.

Phillips, C.J.C., Mohamed, M.O. and Chiy, P.C. (2015) Effects of duration of salt supplementation of sheep on rumen metabolism and the accumulation of elements. *Animal Production Science* 55, 603–610.

Potter, B. J., and McIntosh, G.H. (1974) Effect of salt water ingestion on pregnancy in the ewe and on lamb survival. *Australian Journal of Agricultural Research* 25, 909–917.

Provenza, F.D. (2008) What does it mean to be locally adapted and who cares anyway? *Journal of Animal Science* 86, E271–E284.

Rabelink T.J., and Rotmans, J.I. (2009) Salt is getting under our skin. *Nephrology Dialysis Transplantation* 24, 3282–3283.

Ravagnolo, O., and Misztal, I. (2000) Genetic component of heat stress in dairy cattle, parameter estimation. *Journal of Dairy Science* 83, 2126–2130.

Reece, W.O. (2004) Kidney function in mammals. In: W.O. Reece (Ed.) *Dukes' physiology of domestic animals*, 12th edn. Cornell University Press, Ithaca, New York.

Revell, D.K., Maynard, B., Erkelenz, P.A., and Thomas, D.T. (2015) 'Rangeland Self Herding' – positively influencing grazing distribution to benefit livestock, landscapes and people. *Proceedings of the Australian Rangeland Society Conference.* Alice Springs, 12–17 April 2015.

Rosselle, L., Permentier, L., Verbeke, G., Driessen, B., and Geers, R. (2013) Interactions between climatological variables and sheltering behavior of pastoral beef cattle during sunny weather in a temperate climate. *American Society of Animal Science* 91, 943–949.

Schafflhuber, M., Volpi, N., Dahlmann, A., Hilgers, K.F., Maccari, F., Dietsch, P., Wagner, H., Luft, F.C., Eckardt, K-U., and Titze, J. (2007) Mobilization of osmotically inactive Na+ by growth and dietary salt restriction in rats. *American Journal of Physiology – Renal Physiology* 292, F1490–F1500.

Silanikove, N. (2000) Effects of heat stress on the welfare of extensively managed domestic ruminants. *Livestock Production Science* 67, 1–18.

Squires, V.R. (1976) Walking, watering and grazing behaviour of Merino sheep on two semi-arid rangelands in south-west New South Wales. *The Australian Rangeland Journal* 1, 13–23.

Squires, V.R., and Wilson, A.D. (1971) Distance between food and water supply and its effect on drinking frequency, and food and water intake of Merino and Border Leicester sheep. *Australian Journal of Agricultural Research* 22, 283–290.

Tay, S., Blache, D., Gregg, K., Revell, D.K. (2011) Changes in DNA methylation and renin gene expression in sheep due to in utero salt exposure. *Proceedings of the Federation of Asian and Oceanian Biochemists* and Molecular Biologists (FAOBMB) Conference. Singapore, October 2011.

Tay, S. H., Blache, D., Gregg, K., and Revell, D.K. (2012) Consumption of a high-salt salt diet by ewes during pregnancy alters nephrogenesis in 5-month-old offspring. Animal 6, 1803–1810.

Thomas, D.T., Wilmot, M.G., and Revell, D.K. (2010) Does peer training facilitate adaptation in young cattle relocated from the rangelands to a temperate agricultural grazing system? Proceedings of the 1st Australian and New Zealand Spatially Enabled Livestock Management Symposium pp. 14–15. University of New England, Armidale NSW Australia, 15 July 2010.

Vercoe, J.E., Frisch, J.E., and Moran, J.B. (1972) Apparent digestibility, nitrogen utilization, water metabolism and heat tolerance of Brahman cross, Africander cross and Shorthorn x Hereford steers. Journal of Agricultural Science, Cambridge 79, 71–74.

chapter four

Grazing and animal distribution

Derek W. Bailey

In extensive systems, livestock are typically not fed in small pens and graze rangelands and pastures to obtain food and water. Although many livestock welfare concerns in intensive and extensive systems are similar (e.g., branding, castration, dehorning, and humane slaughter), this chapter will focus on livestock systems that are based on grazing and utilize rangelands and large pastures (> 25 ha). The goal of this discussion is to describe the challenges livestock face while grazing in extensive systems and describe how grazing and livestock management practices can impact animal welfare.

CHALLENGES FOR LIVESTOCK IN EXTENSIVE SYSTEMS

Animals require water, food, and cover to survive and reproduce in a given area, and this arrangement of environmental conditions is defined as habitat (Thomas, 1979). Unlike confined systems and small farms, livestock must locate and consume feed, travel from foraging locations to water, and seek out shelter (cover).

These critical environmental conditions are often spread out and require animals to remember and travel to their locations (Table 4.1). Wildlife face similar challenges, but livestock are usually confined to pastures and must rely on human management to ensure the basic components of habitat are available in their enclosure.

Water

For livestock on extensive rangelands, especially in semi-arid and arid systems, water is critical (see Chapter 3). Cattle on rangelands typically water daily, but periodically may not return to water for over two to three days (Bailey et al., 2010; Stephenson, 2014). Water sources are often limited and fences usually restrict livestock movements, which may prevent animals from reaching alternative water sources (Table 4.1). Cattle often remain within 1.6 km from water and typically do not travel farther than 3.2 km from water (Valentine, 1947). However, Bailey et al. (2010) reported that Brangus cattle regularly traveled up to 4 km from water in the Chihuahuan

New Mexico State University, Las Cruces, USA

Table 4.1 Differences between extensive and intensive systems with respect to habitat for livestock

Habitat characteristic	Intensive – confined operations		Extensive	
	Source	Maintenance	Source	Maintenance
Water	Man-made – tanks and waterers	Check daily and occasional repairs	Man-made – tanks, wells, pipelines, and waterers	Check regularly based on water storage. Repair as necessary
			Man-made – stock dams	Check regularly to make sure water is still available
			Natural – streams, lakes, and springs	
Food	Harvested forage and grain	Provided daily	Rangeland and large pastures	Check forage availability and adjust stocking rates
Shelter and cover	Man-made – barns, shade, and windbreaks	Periodic maintenance	Trees, shrubs, and variable terrain	Minimal, if any

Desert. In drought conditions, limited water availability may restrict stocking levels before forage availability becomes limiting. Springs and stock ponds may dry up during drought, eliminating water sources that are available during normal and favorable precipitation periods. Water requirements vary throughout the year. During the winter, cattle can survive on snow when water sources freeze over (Young and Degen, 1980). Consumption of snow as the sole water source has no apparent ill effect on pregnant beef cows (Degen and Young, 1990). In contrast to winter, water requirements increase during the higher temperatures of summer, especially when animals are lactating (Winchester and Morris, 1956). On extensive systems, there is no greater animal welfare concern for ranchers and livestock producers than ensuring livestock have available water.

Food

In extensive systems, livestock must search out forages rather than consume food from troughs and feeders (Table 4.1). In small pastures, finding forage is not the daunting task that livestock in extensive systems must face. Foraging decisions occur at different spatial and temporal scales in extensive systems and can be organized based on hierarchical theory (Senft et al., 1987). At finer scales, livestock must select among plant species and plant parts and as they move along the grazing pathway. These decisions are made every one or two seconds and are based on nutrient concentration, toxin concentration, and plant size (Bailey et al., 1996). At intermediate scales, livestock regularly select among patches every 1 to 30 minutes. Patches are best defined by the animals when they make a change in their foraging sequence and path and reorient to a new location (Bailey et al., 1996; Jiang and Hudson, 1993), but patches typically vary in size from 0.001 to 1.0 ha (Bailey and Provenza, 2008). Livestock select among patches based on forage quality, forage quantity, social interactions, and to some extent topography. At more coarse scales, livestock select among feeding sites. Feeding sites are defined as the area used during a 1- to 4-hour foraging bout, but cattle often remain near the feeding site selected during the previous evening for their morning feeding bout (Bailey et al., 2004; Low et al., 1981). Correspondingly, feeding sites often include the areas grazed during the evening, following morning, and the area used for resting that night. Criteria likely used for selecting among feeding sites include topography and distance to water as well as factors critical at finer scales such as forage quality, forage quantity, presence of toxins, and secondary compounds. In very extensive systems with large pastures, livestock may select among assemblages of feeding sites, termed as camps (Bailey et al., 1996). Camps are based on a water site (or set of nearby water locations) that animals return to daily (or every few days). Livestock typically move among camps at temporal intervals of weeks to months. At the camp scale, water availability, forage abundance, phenological impacts to forage quality, thermoregulation, and predation are likely the factors used for decision making.

Another major challenge faced by livestock during grazing is that the quality and quantity of forages are constantly changing. Phenological changes and localized variability in precipitation patterns continually affect nutrient concentrations of forages. Grazing by livestock and other herbivores changes quantity as well as quality of forages at a given location. However, livestock and other large herbivores are adapted to this complex and ever-changing environment. Livestock have accurate spatial memories and can remember locations of food (Bailey et al., 1989b; Dumont and Petit, 1998; Laca, 1998). They also can remember the quantity and quality of food found at a location (Bailey et al., 1989a; Bailey and Sims, 1998). Through postingestive feedback they can detect forages that contain high concentrations of nutrients and avoid foods with excessive toxins (Launchbaugh et al., 1993; Provenza, 1995; Provenza, 1996; Villalba et al., 2002). Typically, livestock prefer forages with higher nutrient concentrations and fewer toxins, but animals continually select a mixed diet that changes as relative levels of nutrients and toxin concentration vary. For example, sheep prefer foods with higher levels of protein after consuming a meal high in energy and prefer foods high in energy after consuming a protein-based meal (Villalba and Provenza, 1999). The nutrient intake can also affect livestock's willingness to consume toxins (Villalba and Provenza, 2005). Bailey and Provenza (2008) describe a conceptual model (satiety hypothesis) that summarizes how postingestive feedback of nutrients and toxins affects diet selection and helps livestock maintain a balance between energy- and protein-rich foods. With some modifications, the satiety hypothesis can help explain feeding site selection. Typically, livestock prefer sites with higher forage quality (Bailey, 1995), but periodically they graze in sites with lower forage quality. Livestock can become satiated with a feeding site. In homogeneous areas, cattle regularly alternate among feeding sites (Bailey et al., 1990). On extensive ranches in the western USA, cattle used a different feeding site each day during 70% of the GPS tracking period when the terrain was relatively homogeneous, but in a 9,740 ha pasture enclosing gentle and mountainous terrain cows stayed in the same feeding site for over 10 days during 42% of the tracking period (Bailey et al., 2015).

During periods when vegetation is dormant (e.g., winter or dry season), livestock may not have access to forages that have sufficient nutrient density to meet their metabolic requirements. Dormant forage is typically low in nitrogen, and animals are unable to consume sufficient quantities of crude protein. Unlike wild herbivores, ranchers and land managers often provide supplements to livestock when forage quality is low. Protein supplements improve intake and digestibility of low quality forages (Clanton and Zimmerman, 1970; Schauer et al., 2005).

Shelter and Cover

In extensive systems, livestock must seek out cover and favorable terrain to provide shelter and help maintain homeostasis. During hot weather, livestock seek out shade. For example, cows grazing California rangelands spent 8 hours a day in the shade of trees on sunny summer days (Harris et al., 2002). During the winters, these cows avoided shade trees and grazed southern exposures. Cows sought out favorable terrain that was out of wind during cold and windy

winter weather in Montana (Houseal and Olson, 1995). Similarly, cattle preferred pinyon-juniper woodlands during cold and windy weather. In confined operations, livestock cannot seek out protected terrain and shelter must be provided (Table 4.1) to maintain productivity and some cases homeostasis in cold and wet and especially in cold, wet, and windy weather (Schütz et al., 2010). In contrast, cattle on Montana rangelands did not regularly seek out man-made windbreaks during the winter (Olson et al., 2000). Cattle apparently were adapted to winter weather conditions in Montana and did not benefit greatly from the windbreaks. Access to variable terrain and woodlands allows livestock to seek out shelter during both hot and cold weather.

GRAZING MANAGEMENT IN EXTENSIVE SYSTEMS

In contrast to intensive confined systems where management provides water, food, and shelter for the livestock, animals must seek out water, forage, and shelter in extensive systems. Livestock are adapted to rangeland and pasture systems and typically thrive in such environments, but managers can have powerful and important impacts on livestock habitat, which can affect not only their productivity but their welfare as well. As far as grazing is concerned, managers can really only affect four things (Vallentine, 2001): stocking rate, distribution, season of use, and kind or class of livestock.

Stocking Rate

Stocking rate is the most important decision that ranchers and land managers must make regarding grazing (Briske et al., 2008; Holechek and Galt, 2000). Stocking rate affects livestock productivity and rangeland health (Holechek et al., 1999). For example, Willms et al. (1985) found that heavy stocking rates resulted in severe rangeland degradation, while light stocking levels had no adverse effect on rangeland condition. Hart and Ashby (1998) found that heifer weight gain decreased linearly with increased grazing pressure. Stocking rate changes the competition among animals for forage. At low stocking rates there is little completion for forage but at higher stocking rates, animals compete for forage, especially higher quality forages. In a review, Allison (1985) found that forage intake was directly correlated with forage allowance. At lighter stocking levels where forage allowances are greater, livestock generally consumed more forage each day. However, Pinchak et al. (1990) found that differences between stocking rates for forage intake and diet quality were not consistent throughout the year, but there were periodic differences that they concluded accounted for the variation in cattle performance. With sheep and goats, grazing time, distance traveled, and energy expenditure increased with increasing stocking rates, but no consistent differences in diet quality were detected among stocking rate treatments (Animut et al., 2005). Similarly, Ackerman et al. (2001) found only minor differences in diet quality of calves grazing at different stocking rates, but grazing time increased at higher stocking levels. The authors concluded that the lower gains of calves at higher stocking levels were likely due to increased energy expenditures associated with longer grazing times.

The effects of stocking rate on grazing behavior can be seen over a short time period when stocking densities are high. Olson et al. (1989) found that diet quality and ingestive rate decreased after the first day of a 3-day grazing period in a rotational grazing system. Changing the availability of forage for livestock by manipulating stocking rate and/or stocking density can affect diet quality, forage intake, and performance.

Grazing Distribution

Livestock distribution is another critical factor in grazing management. Cattle and other livestock prefer to graze at sites that contain higher quality forages that are located near water, which is the reason that riparian areas receive a disproportionately large amount of use. Concentration of grazing in riparian areas and other preferred areas can damage riparian areas, degrade wildlife habitat, increase erosion, and reduce water quality (Kauffman and Krueger, 1984; Meehan and Platts, 1978).

Livestock grazing distribution is affected by abiotic factors, such as slope and horizontal and vertical distance to water, and biotic factors such as forage quality and quantity (Bailey et al., 1996). Effects of various terrain attributes cannot be evaluated independently and their cumulative impact should be considered (Bailey, 2005). For example, cattle are much less likely to graze steep slopes that are located over 2 km from water than similar slopes located less than 0.5 km from water. In addition to terrain and forage quality and quantity, weather and social interactions can affect grazing distribution. Uniformity of grazing is usually greater in pastures with homogeneous vegetation but livestock performance may be greater in hetero-geneous areas, especially with lighter stocking levels (Bailey, 2005).

Extensive livestock production systems usually allow livestock to express their natural behavior to a much greater degree than confined systems. However, managers often implement practices to manipulate grazing distribu-tion. Development of new water sources reduces the distance that livestock must travel to water when grazing (Williams, 1954). Strategic supplement placement not only provides nutrients, but can be used to attract cattle to areas that nor-mally are not grazed (Bailey and Welling, 1999). Low-stress herding is a powerful tool to reduce cattle use in riparian areas. Bailey et al. (2008) found that low-stress herding reduced the time cattle spent near streams by about 40%. Herding increased the distance that cows trave-led each day, but there was no indication that the herding had any adverse impacts on weaning weights, cow reproduction, and cow body condition (Bailey, 2014, unpublished results). Grazing distribution practices are likely to have little, if any, adverse impacts on animal welfare.

Season of Use

Changing the season of grazing can be used to give vegetation a rest from defo-liation during the active growing season, which can improve plant vigor and help maintain rangeland health (O'Reagain and Ash, 2002). The periodic rest from defoliation is the basis for rotational graz-ing systems (Vallentine, 2001), but in an extensive review of grazing research, Briske et al. (2008) found little experimen-tal evidence documenting the benefits of

rotational grazing. Other researchers sug-gest that rotational grazing can improve rangeland sustainability and livestock productivity (Teague et al., 2011). In any case, timing of grazing can play a role in animal welfare. During the summer, pastures with trees and woodlands pro-vide shade, which helps livestock cope with high temperatures. In the winter, pastures with variable terrain and wood-lands provide animals the opportunity to get out of the wind. Rugged terrain also allows animals to seek out ridgetops with warmer air temperatures and sunny southern exposures during cold periods without wind (Harris et al., 2002).

Rotational and other grazing systems require periodic movements of live-stock from one pasture to another. Herding animals to another pasture (or paddock) has the potential to adversely affect animal welfare, but in most cases livestock become accustomed to the moves (Petherick, 2005). Low-stress stockmanship techniques developed by Bud Williams can improve the efficiency of herding while minimizing stress to the livestock (Hibbard, 2012a). With training and practice, stockmen using Williams' techniques can readily herd cattle to corrals for husbandry practices (e.g., vaccinations, weaning, and brand-ing), move to new pastures, or relocate cattle from riparian areas to uplands (Bailey et al., 2008; Hibbard, 2012b). In feedlots, Noffsinger and Locatelli (2004) have shown that these low-stress stock-manship techniques can help reduce the incidence of disease. Use of low-stress stockmanship is beneficial practice for mangers who implement various grazing systems and for other rangeland manage-ment practices (Bailey and Stephenson, 2013; Cote, 2004).

Kind and Class of Animal

The kind and class of livestock used in extensive systems is usually based on owner preference, economics, marketing opportunities, terrain, vegetation types, and facilities. However, some livestock species are a better match for certain climates and forage conditions. Sheep and goats are more heat tolerant, require less water and can consume a wider range of forages than cattle (Kay, 1997; Seo, 2008; Seo et al., 2010). Multispecies grazing has several potential benefits, both ecological and economical (Glimp, 1988), but most operations specialize in one species. Use of sheep and goats in extensive operations are often associated with targeted grazing in the USA (Launchbaugh et al., 2006). For example, goats are typically used to consume brush, and their use may increase the abundance of shrubs increases as anticipated (Glimp, 1995). In most arid and semi-arid regions of the world, sheep and goats are more common than cattle (Glimp, 1988). Additional shifts from cattle to sheep and goats may be beneficial in the future if climatic conditions continue to get hotter and drier as projected by many climatologists (Joyce et al., 2013; Polley et al., 2013).

Identification and use of adapted animals in extensive systems is critical for animal welfare in extensive systems (Petherick, 2005). As noted above, livestock must be able to seek out water, food, and shelter in extensive systems, as opposed to being provided these elements of habitat in confined systems (Table 4.1). Livestock evolved on rangelands, but animals are relocated frequently, and they may be unadapted to local conditions. For example, animals that are moved from a familiar to an unfamiliar location suffer more from malnutrition, consumption of poisonous plants, and predation compared to locally adapted animals (Davis and Stamps, 2004; Provenza, 2008). Adaption to local environments is a process that can involve genotype, learning, and experiences and likely genotype and environmental interactions (Provenza, 2008).

Experience early in life plays a critical role in foraging later in life. For example, goats that had experience with blackbrush with their mothers early in life (1 to 4 months of age) consumed 2.5 time more blackbrush (*Coleogyne ramosissima* Torr.), a shrub containing high levels of tannins, compared to similar age naive goats (Distel and Provenza, 1991). Brangus cows that were born and raised in the Chihuahuan Desert in southern New Mexico selected a higher quality diet than naive cows from a subtropical region in east Texas when vegetation was dormant and forage quality was relatively low (Bailey et al., 2010). Diets of locally adapted cows likely contained significant quantities of relatively nutritious shrubs, while the naive cows relied on the less nutritious grasses that they were familiar with. After a period of unusually high levels of precipitation, both local and naive cows selected diets that exceed their requirements, but naive cows tended to select a higher quality diet than locally raised cows (Bailey et al., 2010). Apparently, naive cows focused on the high quality grasses that they were familiar with after extremely favorable precipitation, while local cows continued to mix some shrubs with grasses. Locally adapted livestock have the ability to select higher quality diets than naive animals under typical climatic conditions. In addition, these animals are more likely to consume plants with higher levels of secondary compounds (Provenza, 2008).

Experience and local adaptation also play an important role in grazing distribution. Howery et al. (1996) found that cows showed fidelity to given areas (similar to camps) in an extensive mountain pasture. Cows used the same pasture areas for four consecutive years. These "home range" or camp preferences were affected by their experiences as calves. In a later cross-fostering study, Howery et al. (1998) found that female offspring from the previous study used the same areas as the unrelated cows that they were reared with (foster mothers). In a similar cross-fostering study with two breeds of sheep with different habitat preferences (Key and MacIver, 1980), half of the ewe lambs were reared by another breed. Subsequent observations after weaning showed that cross-fostered lambs showed more fidelity to the habitat type of the breed of ewe that had reared them (foster mothers) rather than their own breed.

In a New Mexico study (Bailey et al., 2010), Brangus cows born and raised in the Chihuahuan Desert were moved to east Texas (a subtropical area) for 3 years and then brought back to New Mexico (termed tourist cows). Cows born and raised in East Texas were brought to the Chihuahuan Desert (termed naive cows). Grazing patterns of naive and tourist cows were compared to cows that were born, raised, and remained in the Chihuahuan Desert (termed native cows). Native cows consistently used areas farther from water than either the naive or tourist cows. However, experience did not affect how far cows traveled

each day. Naive, tourist, and native cows walked the same distance, but native cows spent more time at locations far from water. Native cows also spent less time at water than naive or tourist cows. These results suggest that cows in desert conditions can become adapted and willing to graze areas far from water. After grazing in small temperate pastures in east Texas for 3 years, cows born and raised in desert conditions were apparently no longer adapted.

In addition to experience, genotype also plays an important role in livestock adaptability. For example, *Bos indicus* cattle are more adapted to hot and humid environments than *Bos taurus* cattle (Hoffmann, 2010). Beatty et al. (2006) found that feed intake decreased and water intake increased more for Angus heifers than Brahman heifers in hot and humid conditions. *Bos indicus* cattle are more resistant to heat transfer to the skin than *Bos taurus* (Finch, 1986). Coat color and coat structure play important roles. The thick coats of European breeds provide more insulation and more resistance to heat transfer than the coats of Brahmans, but the slick, dense structure of Brahman coats reflects more solar radiation which helps reduce heat transfer (Finch, 1986). Environmental heat gain is greater for dark-colored cattle than light-colored cattle (Finch et al. 1984), but the impact of coat color and coat structure during high temperatures is greater for European breeds than Brahmans. Increased use of locally heat adapted breeds and Brahman crosses may be warranted based on projected increases in global temperatures (Hoffmann, 2010).

Increased winter survival and abundance of ectoparasites may adversely affect livestock to a greater degree than physiological impacts resulting from anticipated warmer temperatures predicted by climate change models (Karl et al., 2009). In the United States, horn flies (*Huematobia irrituns* [L.]) are a concern for cattle and can reduce gains by 4 to 14% (Byford et al., 1992). Lower rates of gain are at least partially attributed to increased time spent walking and in engaging in avoidance behavior, such as tail switching (Harvey and Launchbaugh, 1982). Ticks (*Amblyomma americanurm* Koch) may have the largest adverse impact on cattle production of all ectoparasites. Ticks can reduce weight gains of British breeds by as much as 30%, but they have very little impact on Brahmans (Byford et al., 1992). Correspondingly, ranchers may need to increase the proportion of Brahman or other insect resistant breeds in their herds if projected increases in ectoparasites occur. White et al. (2003) cautioned that weight losses might be over 1.5 times greater in the future if producers in Australia do not use more adapted cattle (e.g., Brahmans).

In addition to learning and experience, genotype plays a role in diet selection and can be affected by a rancher's breeding program. Winder et al. (1996) found differences in the botanical composition of diets selected by Angus, Hereford, and Brangus cows grazing in the Chihuahuan Desert. In contrast, no important differences in diet selection were detected between Brahman influenced breeds that were adapted to hot, semi-arid conditions, Barzona, Brangus, and Beefmaster (De Alba Becerra et al., 1998). Similarly, Russell et al. (2012) did not find any differences in the quality of diets selected by Angus, Brangus, and Brahman cows. Winder et al. (1995)

reported that the botanical composition of the diet appeared to be heritable, and suggested that part of the differences in diet selection could be attributed to distribution patterns. Cows that traveled farther from water had the opportunity to select more palatable forages that were not available close to water.

Genotype is also important in grazing distribution. Breeds that are adapted to hot temperatures can more readily travel during the summer to forage than unadapted breeds. Herbel and Nelson (1966) observed that Santa Gertrudis cows (Brahman cross) walked further during the spring and summer than Hereford cows. Differences in distance traveled by Herefords and Santa Gertrudis were less pronounced in the fall and winter. Brahman cows walked farther than Angus or Brangus cows, but there was no difference among breeds in the use of areas far from water (Russell et al., 2012). Grazing paths of Brahman cows were more tortuous than Angus or Brangus cows. Cattle breeds developed in mountainous areas tend to graze more uniformly in rough terrain than breeds developed in more gentle topography. Tarentaise cows developed in the French Alps traveled farther vertically from water and tended to use steeper slopes than Hereford cows developed in the relatively gentle rolling terrain of southern England (Bailey et al., 2001). Piedmontese-sired cows developed in the Italian Alps traveled farther from water in mountainous pastures than Angus-sired cows developed in relatively gentle farming areas of eastern Scotland (VanWagoner et al., 2006). Grazing patterns vary among individual cows within a breed. Bailey et al. (2004) documented some of this variation and found that "hill-climber" cows used elevations that were about 50 m higher than "bottom dweller" cows in foothill rangeland pastures in Montana.

Several researchers have suggested that ranchers using mountainous topography might be able to select cows that were more adapted and willing to graze mountainous terrain (Bailey, 2004; Howery et al., 1996; Roath and Krueger, 1982). Ongoing research suggests that genomic selection may provide a practical and affordable means for implementing genetic selection for cattle grazing distribution (Bailey et al., 2013). Genetic markers that are associated with terrain use have been identified. The locations of associated markers are co-located with candidate genes associated with spatial memory, motivation, and locomotion. Genetics tests that are projected to cost about US$30 could be used to develop molecular breeding values for individual cattle. These molecular breeding values could then be used to identify and select bulls and replacement females with favorable genotypes for grazing distribution and to potentially cull animals with unfavorable distribution genotypes. More work needs to be completed to verify these genetic markers and to develop molecular breeding values. However, ongoing work is promising and does not require expensive GPS tracking to estimate breeding values. Selection for grazing distribution is not expected to adversely affect cattle reproduction or calf weaning weights. Two studies showed that there were no consistent phenotypic relationships between terrain use and cow pregnancy rates and cow condition or calf birth dates, birth weights, or weaning weights (Bailey et al., 2001; VanWagoner et al., 2006).

HOW DOES GRAZING AND MANAGEMENT OF EXTENSIVE LIVESTOCK SYSTEMS AFFECT THE FIVE FREEDOMS?

Although implementation of animal welfare into agriculture production in Europe and the United States has differed, the "five freedoms" developed by the Farm Animal Welfare Council (FAWC, 1992) have been used to guide development of standards and management practices (Mench, 2008; Veissier et al., 2008). The first three freedoms have been accepted by livestock producers and can be considered as production traits while the last two freedoms focus on ethological issues (Gonyou, 1994). The five freedoms were developed based on management in confined and intensive livestock production scenarios. Animal welfare concerns in extensive systems differ from intensive systems (Petherick, 2005). The following section will discuss how grazing and livestock management in extensive systems affect the five freedoms (Table 4.2).

Freedom from Thirst, Hunger, and Malnutrition

In most extensive systems, especially arid and semi-arid rangeland, water availability is limited. Ephemeral springs and stock ponds are often used as water sources for livestock. During drought and other hot or dry periods, these water sources may dry up. Man-made wells, pipelines, and tanks can fall into disrepair and fail to provide water. Sufficient wind to pump adequate supplies of water from the windmill may not be available during certain times of the year (e.g., doldrums). Correspondingly, managers

Table 4.2 Application of the "five freedoms of animal welfare" for grazing livestock on extensive systems to ensure proper management

Freedom	Concern	Consequence	Recommended management
1 – from Thirst	No availability of water	Death, heat stress, reduced food intake	Check water sources regularly, store water, move livestock to pasture with water if needed
1 – from Hunger	Lack of forage	Hunger, poor performance, malnutrition	Use appropriate stocking rates, check forage availability, provide supplemental forage, or move livestock when forage is limited
1 – from Malnutrition	Poor forage quality	Malnutrition and poor performance	Provide supplemental protein and minerals when needed, usually when forage is dormant
2 – from Discomfort	Heat stress and cold stress	Ability to maintain thermal balance, poor performance	Ensure pastures have trees or other shade during the summer and trees and/or variable terrain during the winter
3 – from Pain and Injury	Injured or sick animals are not identified and treated	Continued pain and suffering	Check livestock regularly at water during midday to increase probability of seeing all livestock
4 – to Express Normal Behavior	Use of unadapted or inexperienced livestock	Malnutrition, overconsumption of toxins, poor performance, poor grazing distribution	Use adapted breeds and livestock species. Select female replacements from own herd or similar environments. Consider adaptability when selecting sires of female replacements
5 – from Fear and Distress	Animals become panicked and afraid	Stress and poor performance	Use low-stress stockmanship during herding and handling

must check livestock water regularly. The frequency of inspection depends on water capacity, weather, and water demand (number of livestock, kind of livestock, and physiological status). The interval between inspections can be no longer than the time for livestock in the pasture to consume the water that is currently in the tank (drinker). Although frequent monitoring of livestock water requires a substantial labor requirement, the consequences of having no water in a pasture are disastrous. During summer desert conditions, cows can lose over 20% of their body weight when deprived of water for 3 days and will likely perish after 5 days without water (Siebert and Macfarlane, 1975). Recent technological developments may allow ranchers to monitor water availability remotely in extensive pastures using sensors that track water levels in stock tanks and satellite or cell phone technologies to transmit the information to ranch headquarters (Heaton, 2014). These technologies can reduce labor and travel costs and help ensure animal welfare in extensive and rugged pastures.

Livestock can usually roam freely in search of food in extensive systems. Correspondingly, freedom from hunger is primarily determined by stocking rate (Figure 4.1). As stocking rate increases, forage utilization increases and forage intake and diet quality usually decrease (Ackerman et al., 2001; Ash and Stafford Smith, 1996). Weight gain and conception rates decrease as stocking rate increases, but these reductions are usually more rapid in seeded pastures compared to rangeland (Ash and Stafford Smith, 1996; Hart et al., 1988). In extensive rangeland pastures, impacts of stocking rates on livestock production can be masked in the short-term as cattle switch to less preferred forage species and travel farther from water at high stocking rates compared to lower stocking rates (Figure 4.1). However, in the long term, livestock performance at higher stocking levels may decline as rangeland health begins to decline and higher-quality preferred forages are replaced with less desirable species (Ash and Stafford Smith, 1996). Therefore, livestock are more likely to be free from hunger at light stocking levels rather than heavy levels. Ranchers can monitor livestock body condition to determine if livestock are able to obtain enough nutrients from the pasture (Mathis et al., 2002; Selk et al., 1988). As animals lose weight they become thinner, which can be documented by regularly observing body condition. Recent advances in technology make it possible to monitor weight changes on a daily basis while livestock are grazing on pastures and the data could be sent remotely to the ranch headquarters (Brown et al., 2014b). Morris et al. (2012) suggested that monitoring individual sheep weights as they come to water to drink could be used to ensure that animals had sufficient forage and help improve animal welfare. However, monitoring weights of individual sheep can be unreliable because of animal identification issues and regularly weighing groups of sheep may be more repeatable (Brown et al., 2014a; Brown et al., 2012). Systems similar to those used for sheep are being tested for beef cattle grazing extensive rangeland pastures (Leigo et al., 2012). If livestock are becoming thinner (lower body condition scores or losing weight), stocking rates may be too high for the forage available. In such cases, livestock can be moved

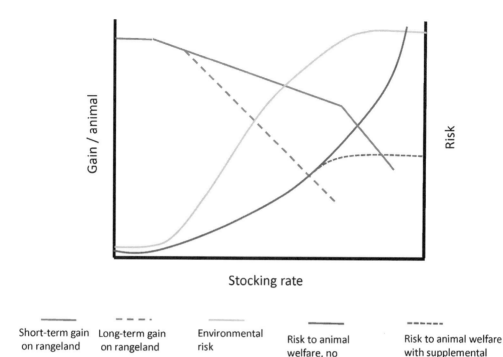

Short-term gain   Long-term gain   Environmental   Risk to animal   Risk to animal welfare
on rangeland   on rangeland   risk   welfare, no   with supplemental
                                                       supplemental feeding   feeding

Figure 4.1: Conceptual diagram of expected changes in livestock gain on rangelands over the short and long term and risks associated with environmental condition and animal welfare with increasing stocking rates. Expected long-term changes in livestock gain assume that rangeland health and productivity decline over time with heavier stocking rates. Risk associated with animal welfare at higher stocking rates can be ameliorated with supplemental feeding. Adapted from Ash and Stafford Smith (1996) and Reece et al. (2001).

to a new pasture and/or animals can be gathered and sold or moved to a new property or ranch with available forage. Livestock can also be fed harvested forage and grains as a supplement, or more likely a substitute for rangeland forage, which should allow livestock to maintain body condition and alleviate concerns with hunger and malnutrition (Figure 4.1).

Livestock can also become thin and lose weight in extensive systems if forage quality is low. In such cases, animals cannot consume sufficient forage to meet nutrient requirements because the concentration of nutrients is too low and/ or the forage cannot be readily digested

(Clanton and Zimmerman, 1970; Leng, 1990). As grasses and other forages become dormant, crude protein levels decline and fiber levels increase (Murillo-Ortiz et al., 2014; Van Soest, 1994). When forage quality is low, ruminant livestock can suffer malnutrition even if there is an abundance of forage and stocking rates are light. Providing livestock supplemental feeds that are high in crude protein can improve forage digestibility and intake when forage quality is low (Clanton and Zimmerman, 1970; Schauer et al., 2005; Titgemeyer et al., 2004). Correspondingly, ranchers should provide supplemental protein when forage quality is low to maintain livestock performance and welfare.

Managers can evaluate forage quality of their pastures in a variety of ways. The simplest and least accurate approach is to monitor plant phenology. When grasses mature and become yellow in color and are no longer green, crude protein concentration may be below livestock requirements. Collecting forage samples and sending them to a laboratory for nutritional analysis provides an accurate assessment of the nutrient content of the sample. The issue is that it is difficult to determine the combination of forages that livestock consume. Researchers often use cannulated or fistulated animals to collect diet samples for nutritional and botanical composition analyses (Holechek et al., 1982), which is not practical for livestock producers in extensive systems. Near infrared reflectance spectroscopy (NIRS) analyses of fecal samples can be used to estimate the quality of diets selected by livestock (Li et al., 2007; Stuth et al. 2003). Periodic collection of fecal samples and NIRS analyses is a practical tool for managers to monitor diet quality of their livestock (Landau et al. 2006; Walker, 2010). This technology can also be used to determine if the level of supplementation provided by management is allowing livestock to maintain an adequate diet (Dixon and Coates, 2010).

Livestock can also become malnourished on extensive systems from mineral deficiencies and mineral supplementation may be required to maintain livestock health and productivity. In the western United States, mineral deficiencies are usually an issue when forages are dormant (e.g., winter or drought), but in other areas soils may be deficient in phosphorus and other minerals and mineral supplementation may be required throughout the year

(Holechek and Herbel, 1986). In areas where phosphorus is deficient in the soil, fertilization of pastures with phosphorus can improve forage quality for several years (Black and Wight, 1979).

Endoparasites and ectoparasites impact livestock health and productivity. For example, weight gains of young cattle with *Helminth* infections can be 15 to 30% less than uninfected animals (Craig, 1988). Horn flies can not only reduce animal gains but increase heart rate, respiration rate, and water consumption (Byford et al., 1992). Brahman cattle are relatively resistant (White et al., 2003), but gains of British and other *Bos taurus* breeds can be reduced by over 30% from tick infestations (Byford et al. 1992). When parasite loads become excessive, managers must treat their herds or animal productivity will decline and animal health may be compromised. Failure to treat livestock for endo- and ectoparasites can adversely affect animal welfare.

Freedom from Discomfort

Livestock must find shelter in most extensive systems. However, managers should recognize seasons that livestock may need shelter and ensure the pastures where they graze can provide any needed shelter. During the summer, pastures containing trees and northern aspects may be beneficial during hot summer periods (Bear et al., 2012). Shade can reduce an animal's radiant heat load by 30% (Blackshaw and Blackshaw, 1994). Providing shade to steers in Oklahoma during the summer improved gain by 6% (McIlvain and Shoop, 1971). In the winter, livestock can use variable terrain and forested areas to find favorable microclimates and avoid high winds and cold temperatures that contribute to cold stress (Houseal and Olson, 1995). Harris et al. (2002) found that cattle took advantage of rugged terrain to help maintain thermal balance during the winter. Cattle grazed southern exposures during the day and spent the night on warmer ridges. During cold and windy weather, cattle sought out juniper woodlands in New Mexico (Black Rubio et al., 2008).

Freedom from Pain, Injury and Disease

Monitoring livestock health is more difficult in extensive systems than in confinement operations and small farms. Livestock are spread out in extensive systems. Consequently, ranchers usually travel to water sources at midday in arid and semi-arid environments to find, observe, and monitor health and condition of livestock. Livestock usually travel to water at midday and graze during the early morning and evening (Bailey et al., 2004; Gregorini, 2012; Low et al., 1981). Farmers in the Italian Alps place salt in accessible areas in gentle terrain to help gather cattle for periodic herd counts and monitoring cattle health (personal communication Marco Pittarello, 2013, Turino, Italy). If an animal is found to be injured or sick, treatment of the animal is more difficult in extensive systems. Many producers herd sick or injured animal to pens and corrals for treatment, which can prove difficult especially if the animal is lame or very ill. Some ranchers in the western United States rope cattle to constrain them for treatment. However, roping and treating cattle in extensive systems requires skill and can be hazardous for the animal and workers.

Freedom to Express Normal Behavior

In extensive systems, livestock are free-roaming and have the opportunity to express normal behavior. However, animals who are not adapted to local conditions may not be able to behave normally. For example, British cattle breeds are less heat tolerant than the Brahman cattle, and they spend more time at water and do not travel as far as cattle with Brahman breeding during hot temperatures (Herbel and Nelson, 1966; Moran, 1973). Cattle with recent experience in desert conditions are able to graze farther from water than naive animals or animals that do not have recent experience (Bailey et al., 2010). Young and relatively inexperienced cows were not as able to identify and use protected microclimates in cold winter conditions in Montana as older and more experienced cows (Beaver and Olson, 1997). Young livestock likely learn about foraging and survival from their mothers, which makes early experience beneficial (Distel and Provenza, 1991; Launchbaugh and Howery, 2005; Provenza, 2008). Correspondingly, ranchers should try to select replacement females from their own herd. If female replacements must be purchased, producers should attempt to select females from environments that are similar to those at their ranches. Selection of males in extensive livestock operations is also important because the selection differential for males is much greater than for females and can result in greater genetic progress than female selection (Falconer, 1981). However, males do not contribute to early learning of livestock in most operations, thus their contribution is from their genotype. Producers should consider if breeding males have genotypes that adapted to the local environment (Hohenboken et al., 2005; Mignon-Grasteau et al., 2005). At a minimum, breeding males should not have genotypes that will make replacement females less adapted to local conditions. With terminal sire breeding programs, bulls and rams can have traits that may not be desirable for local conditions because all offspring are sold and not used as replacements (Gregory and Cundiff, 1980). Livestock producers should make adaptability to local conditions a priority in the breeding and female replacement management programs to enhance animal welfare in extensive systems.

Freedom from Fear and Distress

Most of the time livestock in extensive systems are free from fear and distress because they are free roaming. However, managers must periodically gather livestock and move them to another pasture or to corrals for husbandry practices such as branding, weaning, pregnancy detection, or calving. Livestock can become afraid and distressed when they are herded and handled, especially if inappropriate stockmanship skills are applied (Grandin, 1997). The lack of regular human contact in extensive systems can contribute to livestock distress during herding and handling. Low-stress stockmanship techniques are an effective tool to reduce livestock stress during herding and handling (Cote, 2004; Hibbard, 2012a; Hibbard, 2012b). Appropriately designed corrals and livestock handling facilities can also help reduce animal stress (Grandin, 1980, 1990; Hibbard and Locatelli, 2014). Training and education of ranchers and livestock workers on low-stress stockmanship and recognition of the stress that can occur with inappropriate handling will help ensure that livestock are free from fear and distress.

Livestock may also face predation rangelands. On US rangelands, sheep and goat face predation from coyotes and on some rangelands cattle and other livestock must deal with predation from mountain lions and wolves. These predators may increase the level of fear and distress in livestock (see Chapter 6).

WHAT ELSE DO WE NEED TO KNOW ABOUT LIVESTOCK WELFARE AND GRAZING IN EXTENSIVE SYSTEMS?

Although there is always more to learn, there is a tremendous amount of research addressing stocking rates, livestock supplementation on rangelands, cold and heat stress, and treatment of animal disease. These issues affect livestock productivity in extensive systems and have been the focus of animal science and rangeland management scientists for decades. In contrast, livestock adaptability has received much less attention by researchers and livestock breeders (Hohenboken et al., 2005; Petherick, 2005). Learning may help livestock cope with their ever-changing environments. Broom (2008) describes animal welfare as an animal's state as it attempts to cope with its environment. Correspondingly, adaption is a critical factor to improve animal welfare. Adaption can occur through learning and experience and through natural and artificial selection. The ability for animals to learn and adapt is beginning to be addressed by researchers (Provenza, 2008), but many questions remain. For example, Howery et al. (1998) demonstrated that young cows learn to graze in the same areas as their mothers, but Wesley et al. (2008) observed that Angus cows were at least 100 m from their calves for 4 to 15 hours per day. With this amount of separation, the method by which calves learn where to graze from their mother needs more investigation. Recent advancements in molecular genetics allows researchers to investigate the role that genetics plays in affecting traits that are difficult to measure, such as adaptability and grazing distribution. The genotype can be assessed directly rather than inferred from phenotypic measures and pedigrees, which allows researchers to use genome-wide association studies to examine the relationships between phenotypic traits and genetic markers (Eggen, 2012).

Another exciting development in animal welfare for livestock production in extensive systems is low-stress stockmanship developed by Bud Williams (Hibbard, 2012a). These techniques have been successfully used to reduce time livestock spend in riparian areas (Bailey et al., 2008) and for focusing cattle grazing for specific land management objectives such as fine fuel reduction and habitat improvement (Bruegger, 2012; Pollak, 2007; Stephenson, 2014). Currently, we need more research evaluating the extent to which low-stress stockmanship can reduce livestock stress during herding and handling and additional studies examining the efficacy of using these techniques for targeting grazing to achieve land management goals and enhance ecosystem services from extensive rangelands.

Advances in technology provide real potential to enhance animal welfare in extensive systems. The ability to remotely monitor water levels in tanks in extensive systems could reduce, not only labor and travel costs, but also the probability that livestock will not have water. Problems with the water supply could be detected almost immediately if sensor and communication issues can be worked out. Walk-over weighing technologies have the potential to identify individual animals that are losing weight (or failing to gain weight at expected levels), which might be indicative of illness or malnutrition. Additional refinement of sensors used to identify individual animals and development of equipment used to ensure weights reflect individual animals is needed. Real-time animal tracking can be used to identify parturition (Dobos et al., 2014) and potentially other behaviors, but additional research and technology development is necessary (Leigo et al., 2012).

CONCLUSIONS AND RECOMMENDATIONS

The number one concern for ranchers and land managers regarding animal welfare in extensive systems is to ensure that livestock have available water. Most extensive systems are located in arid and semi-arid areas and livestock can suffer and die from thirst in only a few days if they cannot find water. The potential for hunger and malnutrition is greater when managers use heavy stocking rates. Livestock on extensive rangelands can likely survive short term periods of heavy stocking levels, but the risk of environmental degradation and potential adverse consequences on forage intake and productivity suggest that light to moderate stocking rates will help maintain animal welfare. Variability of weather, vegetation, and terrain require that livestock be adapted to local conditions to perform and behave normally. Ranchers should try to select female replacement animals from their herd or from livestock operations with similar environmental conditions. Ranchers should consider adaptability when selecting sires of female replacements. Finally, incorporation of low-stress stockmanship into extensive ranching operations should not only improve animal welfare, but improve the effectiveness and stress levels of livestock handlers.

REFERENCES

Ackerman, C., Purvis, H., Horn, G., Paisley, S., Reuter, R., and Bodine, T. (2001) Performance of light vs heavy steers grazing Plains Old World bluestem at three stocking rates. *Journal of Animal Science* 79, 493–499.

Allison, C. (1985) Factors affecting forage intake by range ruminants: a review. *Journal of Range Management*, 305–311.

Animut, G., Goetsch, A., Aiken, G., Puchala, R., Detweiler, G., Krehbiel, C., Merkel, R., Sahlu, T., Dawson, L., and Johnson, Z. (2005) Grazing behavior and energy expenditure by sheep and goats co-grazing grass/forb pastures at three stocking rates. *Small Ruminant Research* 59, 191– 201.

Ash, A.J., and Stafford Smith, D.M. (1996) Evaluating stocking rate impacts in rangelands: animals don't practice what we preach. *The Rangeland Journal* 18, 216–243.

Bailey, D. (1995) Daily selection of feeding areas by cattle in homogeneous and heterogeneous environments. *Applied Animal Behaviour Science* 45, 183–200.

Bailey, D., Rittenhouse, L., Hart, R., Swift, D., and Richards, R. (1989a) Association of relative food availabilities and locations by cattle. *Journal of Range Management* 42, 480–482.

Bailey, D.W. (2004) Management strategies for optimal grazing distribution and use of arid rangelands. *Journal of Animal Science* 82(13 suppl), E147–E153.

Bailey, D.W. (2005) Identification and creation of optimum habitat conditions for livestock. *Rangeland Ecology & Management* 58, 109–118.

Bailey, D.W., Gross, J.E., Laca, E.A., Rittenhouse, L.R., Coughenour, M.B., Swift, D.M., and Sims, P.L. (1996) Mechanisms that result in large herbivore grazing distribution patterns. *Journal of Range Management* 49, 386–400.

Bailey, D.W., Keil, M.R., and Rittenhouse, L.R. (2004) Research observation: Daily movement patterns of hill climbing and bottom dwelling cows. *Journal of Range Management* 57, 20–28.

Bailey, D.W., Kress, D.D., Anderson, D.C., Boss, D.L., and Miller, E.T. (2001) Relationship between terrain use and performance of beef cows grazing foothill rangeland. *Journal of Animal Science* 79, 1883–1891.

Bailey, D.W., and Provenza, F.D. (2008) Mechanisms determining large-herbivore distribution. In: Prins, H.T.T. and Van Langevelde, F. (Eds.) *Resource ecology: Spatial and temporal dynamics of foraging.* Springer, Dordrecht, Netherlands. pp. 7–28.

Bailey, D.W., Rittenhouse, L.R., Hart, R.H., and Richards, R.W. (1989b) Characteristics of spatial memory in cattle. *Applied Animal Behaviour Science* 23, 331–340.

Bailey, D.W., and Sims, P.L. (1998) Association of food quality and locations by cattle. *Journal of Range Management* 51, 2–8.

Bailey, D.W., and Stephenson, M. (2013) Integrating stockmanship into rangeland management. *Stockmanship Journal* 2(1), 1–12.

Bailey, D.W., Stephenson, M., Thomas, M.G., Medrano, J.F., Rincon, G., Cánovas, A., Lunt, S., and Lipka, A. (2013) Manipulation of the spatial grazing behaviour of cattle in extensive and mountainous rangelands. Proceedings of the 17th International Meeting of the FAO-CIHEAM Mountain Pasture Network, June 5–7, 2013. FAO-CiHEAM Trivero, Italy.

Bailey, D.W., Stephenson, M.B., and Pittarello, M. (2015) Effect of terrain heterogeneity on feeding site selection and livestock movement patterns. *Animal Production Science* (in press).

Bailey, D.W., Thomas, M.G., Walker, J.W., Witmore, B.K., and Tolleson, D. (2010) Effect of previous experience on grazing patterns and diet selection of Brangus cows in the Chihuahuan Desert. *Rangeland Ecology & Management* 63, 223–232.

Bailey, D.W., VanWagoner, H.C., Weinmeister, R., and Jensen, D. (2008) Evaluation of low-stress herding and supplement placement for managing cattle grazing in riparian and upland areas. *Rangeland Ecology & Management* 61, 26–37.

Bailey, D.W., Walker, J.W., and Rittenhouse, L.R. (1990) Sequential analysis of cattle location: day-to-day movement patterns. *Applied Animal Behaviour Science* 25, 137–148.

Bailey, D.W., and Welling, G.R. (1999) Modification of cattle grazing distribution with dehydrated molasses supplement. *Journal of Range Management* 52, 575–582.

Bear, D.A., Russell, J.R., and Morrical, D.G. (2012) Physical characteristics, shade distribution, and tall fescue effects on cow temporal/spatial distribution in Midwestern pastures. *Rangeland Ecology & Management* 65, 401–408.

Beatty, D.T., Barnes, A., Taylor, E., Pethick, D., and McCarthy, M. (2006) Physiological responses of Bos taurus and Bos indicus cattle to prolonged, continuous heat and humidity. *Journal of Animal Science* 84, 972–985.

Beaver, J., and Olson, B. (1997) Winter range use by cattle of different ages in southwestern Montana. *Applied Animal Behaviour Science* 51, 1–13.

Black, A., and Wight, J.R. (1979) Range fertilization: Nitrogen and phosphorus uptake and recovery over time. *Journal of Range Management* 32, 349–353.

Black Rubio, C.M., Cibils, A.F., Endecott, R.L., Petersen, M.K., and Boykin, K.G. (2008) Piñon-juniper woodland use by cattle in relation to weather and animal reproductive state. *Rangeland Ecology & Management* 61, 394–404.

Blackshaw, J.K., and Blackshaw, A. (1994) Heat stress in cattle and the effect of shade on production and behaviour: A review. *Animal Production Science* 34, 285–295.

Briske, D.D., Derner, J.D., Brown, J.R., Fuhlendorf, S.D., Teague, W.R., Havstad, K.M., Gillen, R.L., Ash, A.J., and Willms, W.D. (2008) Rotational grazing on rangelands: Reconciliation of perception and experimental evidence. *Rangeland Ecology & Management* 61, 3–17.

Broom, D. M. 2008. Welfare assessment and relevant ethical decisions: key concepts. *Annual Review of Biomedical Sciences* 10, T79–T90.

Brown, D., Savage, D., and Hinch, G. (2014a) Repeatability and frequency of in-paddock sheep walk-over weights: Implications for individual animal management. *Animal Production Science* 54, 207–213.

Brown, D., Savage, D., Hinch, G., and Hatcher, S. (2014b) Monitoring liveweight in sheep is a valuable management strategy: A review of available technologies. *Animal Production Science* (in press).

Brown, D., Savage, D., Hinch, G., and Semple, S. (2012) Mob-based walk-over weights: similar to the average of individual static weights? *Animal Production Science* 52, 613–618.

Bruegger, R. (2012) Use of targeted grazing in Arizona to accomplish rangeland management goals and herder observations of indicators and causal factors influencing rangeland change in Mongolia. MS Thesis, University of Arizona, Tucson, AZ.

Byford, R.L., Craig, M.E., and Crosby, B.L. (1992) A review of ectoparasites and their effect on cattle production. *Journal of Animal Science* 70, 597–602.

Clanton, D., and Zimmerman, D. (1970) Symposium on pasture methods for maximum production in beef cattle: Protein and energy requirements for female beef cattle. *Journal of Animal Science* 30, 122–132.

Cote, S. (2004) *Stockmanship: A powerful tool for grazing lands management.* USDA Natural Resources Conservation Service, Boise, ID.

Craig, T. M. (1988) Impact of internal parasites on beef cattle. *Journal of Animal Science* 66, 1565–1569.

Davis, J.M., and Stamps, J.A. (2004) The effect of natal experience on habitat preferences. *Trends in Ecology & Evolution* 19, 411–416.

De Alba Becerra, R., Winder, J., Holechek, J., and Cardenas, M. (1998) Diets of 3 cattle breeds on Chihuahuan Desert rangeland. *Journal of Range Management 51*, 270–275.

Degen, A., and Young, B. (1990) The performance of pregnant beef cows relying on snow as a water source. *Canadian Journal of Animal Science* 70, 507–515.

Distel, R.A., and Provenza, F.D. (1991) Experience early in life affects voluntary intake of blackbrush by goats. *Journal of Chemical Ecology* 17, 431–450.

Dixon, R.M., and Coates, D.B. (2010) Diet quality estimated with faecal near infrared reflectance spectroscopy and responses to N supplementation by cattle grazing buffel grass pastures. *Animal Feed Science and Technology* 158, 115–125.

Dobos, R.C., Dickson, S., Bailey, D.W., and Trotter, M.G. (2014) The use of GNSS technology to identify lambing behaviour in pregnant grazing Merino ewes. *Animal Production Science* 54, 1722–1727.

Dumont, B., and Petit, M. (1998) Spatial memory of sheep at pasture. *Applied Animal Behaviour Science* 60, 43–53.

Eggen, A. (2012) The development and application of genomic selection as a new breeding paradigm. *Animal Frontiers* 2, 10–15.

Falconer, D.S. (1981) *Introduction to quantitative genetics.* Longman, London, UK.

Farm Animal Welfare Council (1992) FAWC updates the five freedoms. *Veterinary Record* 131, 357.

Finch, V. (1986) Body temperature in beef cattle: Its control and relevance to production in the tropics. *Journal of Animal Science* 62, 531–542.

Finch, V.A., Bennett, I.L., and Holmes, C.R. (1984) Coat colour in cattle: effect on thermal balance, behaviour and growth, and relationship with coat type. *The Journal of Agricultural Science* 102, 141–147.

Glimp, H.A. (1988) Multi-species grazing and marketing. *Rangelands* 10(6), 275–278.

Glimp, H.A. (1995) Meat goat production and marketing. *Journal of Animal Science* 73, 291–295.

Gonyou, H.W. (1994) Why the study of animal behavior is associated with the animal welfare issue. *Journal of Animal Science* 72, 2171–2177.

Grandin, T. (1980) Observations of cattle behavior applied to the design of cattle-handling facilities. *Applied Animal Ethology* 6, 19–31.

Grandin, T. (1990) Design of loading facilities and holding pens. *Applied Animal Behaviour Science* 28, 187–201.

Grandin, T. (1997) Assessment of stress during handling and transport. *Journal of Animal Science* 75, 249–257.

Gregorini, P. (2012) Diurnal grazing pattern: Its physiological basis and strategic management. *Animal Production Science* 52, 416–430.

Gregory, K., and Cundiff, L. (1980) Crossbreeding in beef cattle: Evaluation of systems. *Journal of Animal Science* 51, 1224–1242.

Harris, N.R., Johnson, D.E., George, M.R., and McDougald, N.K. (2002) *The effect of topography, vegetation, and weather on cattle distribution at the San Joaquin experimental range, California.* USDA Forest Service Technical Report PSW-GTR-184. USDA Forest Service, Albany, CA.

Hart, R., Waggoner Jr, J., Dunn, T., Kaltenbach, C., and Adams, L. (1988) Optimal stocking rate for cow-calf enterprises on native range and complementary improved pastures. *Journal of Range Management* 41, 435–441.

Hart, R.H., and Ashby, M.M. (1998) Grazing intensities, vegetation, and heifer gains: 55 years on shortgrass. *Journal of Range Management* 51, 392–398.

Harvey, T.L., and Launchbaugh, J.L. (1982) Effect of horn flies (Diptera, muscidae) on behavior of cattle. *Journal of Economic Entomology* 75, 25–27.

Heaton, K. (2014) *Save time and money: Using remote sensing technology to monitor stock tanks.* Utah State University Extension, Logan, UT.

Herbel, C.H., and Nelson, F.N. (1966) Activities of Hereford and Santa Gertrudis cattle on southern New Mexico range. *Journal of Range Management* 19, 173–176.

Hibbard, W. (2012a) Bud Williams' low stress livestock handling. *Stockmanship Journal* 1(1), 6–163.

Hibbard, W. (2012b) Low-stress livestock handling: Mapping the territory. *Stockmanship Journal* 1(2), 10–30.

Hibbard, W., and Locatelli, L. (2014) Grandin's approach to facilities and animal handling: An analysis. *Stockmanship Journal* 3(1), 1–23.

Hoffmann, I. (2010) Climate change and the characterization, breeding and conservation of animal genetic resources. *Animal Genetics* 41, 32–46.

Hohenboken, W., Jenkins, T., Pollak, J., Bullock, D., and Radakovich, S. (2005) Genetic improvement of beef cattle adaptation in America. In: *Proceedings of the Beef Improvement Federation's 37th Annual Research Symposium and Annual Meeting.* Beef Improvement Federation, Brookings, SD. pp. 115–120.

Holechek, J., Vavra, M., and Pieper, R. (1982) Methods for determining the nutritive quality of range ruminant diets: A review. *Journal of Animal Science* 54, 363–376.

Holechek, J.L., and Galt, D. (2000) Grazing intensity guidelines. *Rangelands* 22(3), 11–14.

Holechek, J.L., Gomez, H., Molinar, F., and Galt, D. (1999) Grazing studies: What we've learned. *Rangelands* 21(2), 12–16.

Holechek, J.L., and Herbel, C.H. (1986) Supplementing range livestock. *Rangelands* 8(1), 29–33.

Holechek, J.L., Vavra, M., and Pieper, R.D. (1982) Botanical composition determination of range herbivore diets: A review. *Journal of Range Management* 35, 309–315.

Houseal, G., and Olson, B. (1995) Cattle use of microclimates on a northern latitude winter range. *Canadian Journal of Animal Science* 75, 501–507.

Howery, L.D., Provenza, F.D., Banner, R.E., and Scott, C.B. (1996) Differences in home range and habitat use among individuals in a cattle herd. *Applied Animal Behaviour Science* 49, 305–320.

Howery, L.D., Provenza, F.D., Banner, R.E., and Scott, C.B. (1998) Social and environmental factors influence cattle distribution on rangeland. *Applied Animal Behaviour Science* 55, 231–244.

Jiang, Z., and Hudson, R.J. (1993) Optimal grazing of wapiti (Cervus elaphus) on grassland: Patch and feeding station departure rules. *Evolutionary Ecology* 7, 488–498.

Joyce, L.A., Briske, D.D., Brown, J.R., Polley, H.W., McCarl, B.A., and Bailey, D.W. (2013) Climate change and North American rangelands: Assessment of mitigation and adaptation strategies. *Rangeland Ecology & Management* 66, 512–528.

Karl, T.R., Melillo, J.M., and Peterson, T.C. (2009) *Global climate change impacts in the United States.* Cambridge University Press, Cambridge, UK.

Kauffman, J.B., and Krueger, W.C. (1984) Livestock impacts on riparian ecosystems and streamside management implications. A review. *Journal of Range Management* 37, 430–438.

Kay, R.N.B. (1997) Responses of African livestock and wild herbivores to drought. *Journal of Arid Environments* 37, 683–694.

Key, C., and MacIver, R. (1980) The effects of maternal influences on sheep: Breed differences in grazing, resting and courtship behaviour. *Applied Animal Ethology* 6, 33–48.

Laca, E.A. (1998) Spatial memory and food searching mechanisms of cattle. *Journal of Range Management* 51, 370–378.

Landau, S., Glasser, T., and Dvash, L. (2006) Monitoring nutrition in small ruminants with the aid of near infrared reflectance spectroscopy (NIRS) technology: A review. *Small Ruminant Research* 61, 1–11.

Launchbaugh, K., Provenza, F.D., and Burritt, E.A. (1993) How herbivores track variable environments: Response to variability of phytotoxins. *Journal of Chemical Ecology* 19, 1047–1056.

Launchbaugh, K., Walker, J. (2006) Targeted grazing—a new paradigm for livestock management. In: *Targeted grazing: A natural approach to vegetation management and landscape enhancement.* American Sheep Industry Association, Centennial, CO, USA. pp. 2–8.

Launchbaugh, K.L., and Howery, L.D. (2005) Understanding landscape use patterns of livestock as a consequence of foraging behavior. *Rangeland Ecology & Management* 58, 99–108.

Leigo, S., Brennan, G., Beutel, T., Phelps, D., Driver, T., and Trotter, M.G. (2012) Overview of technology products for the beef industry of remote Australia. In: CRC-REP Working Paper CW009. Ninti One Limited, Alice Springs, NT Australia.

Leng, R. (1990) Factors affecting the utilization of "poor-quality" forages by ruminants particularly under tropical conditions. *Nutrition Research Reviews* 3, 277–303.

Li, H., Tolleson, D., Stuth, J., Bai, K., Mo, F., and Kronberg, S. (2007) Faecal near infrared reflectance spectroscopy to predict diet quality for sheep. *Small Ruminant Research* 68, 263–268.

Low, W., Tweedie, R., Edwards, C., Hodder, R., Malafant, K., and Cunningham, R. (1981) The influence of environment on daily maintenance behaviour of free-ranging Shorthorn cows in central Australia. I. General introduction and descriptive analysis of day-long activities. *Applied Animal Ethology* 7, 11–26.

Mathis, C.P., Sawyer, J.E., and Parker, R. (2002) *Managing and feeding beef cows using body condition scores.* New Mexico State University, Las Cruces, NM.

McIlvain, E.H., and Shoop, M.C. (1971) Shade for improving cattle gains and rangeland use. *Journal of Range Management* 24, 181–184.

Meehan, W.R., and Platts, W.S. (1978) Livestock grazing and the aquatic environment. *Journal of Soil and Water Conservation* 33, 274–278.

Mench, J.A. (2008) Farm animal welfare in the USA: Farming practices, research, education, regulation, and assurance programs. *Applied Animal Behaviour Science* 113, 298–312.

Mignon-Grasteau, S., Boissy, A., Bouix, J., Faure, J.-M., Fisher, A.D., Hinch, G.N., Jensen, P., Le Neindre, P., Mormède, P., and Prunet, P. (2005) Genetics of adaptation and domestication in livestock. *Livestock Production Science* 93, 3–14.

Moran, J. (1973) Heat tolerance of Brahman cross, buffalo, Banteng and Shorthorn steers during exposure to sun and as a result of exercise. *Australian Journal of Agricultural Research* 24, 775–782.

Morris, J.E., Cronin, G.M., and Bush, R.D. (2012) Improving sheep production and welfare in extensive systems through precision sheep management. *Animal Production Science* 52, 665–670.

Murillo-Ortiz, M., Mellado-Bosque, M., Herrera-Torres, E., Reyes-Estrada, O., and Carrete-Carreón, F.O. (2014) Seasonal diet quality and metabolic profiles of steers grazing on Chihuahuan desert rangeland. *Livestock Science* 165, 61–65.

Noffsinger, T., and Locatelli, L. (2004) Low-stress cattle handling: An overlooked dimension of management. In: *Proceedings of the Meeting of the Academy of Veterinary Consultants* 32.2, 65–78.

O'Reagain, P., and Ash, A. (2002) Principles of sustainable grazing management for the northern savannas. In: Nicholson, S. and Wilcox, D. (Eds.) *Proceedings of the 12th Biennial Conference of the Australian Rangeland Society*, Australian Rangeland Society, Kalgoorie, WA, Australia. pp. 2–5.

Olson, B., Wallander, R., and Paterson, J. (2000) Do windbreaks minimize stress on cattle grazing foothill winter range? *Canadian Journal of Animal Science* 80, 265–272.

Olson, K.C., Rouse, G.B., and Malechek, J.C. (1989) Cattle nutrition and grazing behavior during short-duration-grazing periods on crested wheatgrass range. *Journal of Range Management* 42, 153–158.

Petherick, J.C. (2005) Animal welfare issues associated with extensive livestock production: The northern Australian beef cattle industry. *Applied Animal Behaviour Science* 92, 211–234.

Pinchak, W.E., Canon, S.K., Heitschmidt, R.K., and Dowher, S.L. (1990) Effect of long-term, year-long grazing at moderate and heavy rates of stocking on diet selection and forage intake dynamics. *Journal of Range Management* 43, 304–309.

Pollak, E.R. (2007) Evaluation of low-stress herding and supplement placement to modify cattle distribution and improve pronghorn habitat. MS Thesis, New Mexico State University, Las Cruces, NM.

Polley, H.W., Briske, D.D., Morgan, J.A., Wolter, K., Bailey, D.W., and Brown, J.R. (2013) Climate change and North American rangelands: Trends, projections, and implications. *Rangeland Ecology & Management* 66, 493–511.

Provenza, F.D. (1995) Postingestive feedback as an elementary determinant of food preference and intake in ruminants. *Journal of Range Management* 48, 2–17.

Provenza, F.D. (1996) Acquired aversions as the basis for varied diets of ruminants foraging on rangelands. *Journal of Animal Science* 74, 2010–2020.

Provenza, F.D. (2008) What does it mean to be locally adapted and who cares anyway? *Journal of Animal Science* 86(14 suppl): E271–E284.

Reece, P.E., Volesky, J.D., and Schacht, W.H. (2001) *Integrating management objectives and grazing strategies on semi-arid rangeland, EC158*. University of Nebraska Cooperative Extension, Lincoln, NE.

Roath, L.R., and Krueger, W.C. (1982) Cattle grazing and behavior on a forested range. *Journal of Range Management* 35, 332–338.

Russell, M.L., Bailey, D.W., Thomas, M.G., and Witmore, B.K. (2012) Grazing distribution and diet quality of Angus, Brangus, and Brahman cows in the Chihuahuan Desert. *Rangeland Ecology & Management* 65, 371–381.

Schauer, C.S., Bohnert, D.W., Ganskopp, D.C., Richards, C.J., and Falck, S.J. (2005) Influence of protein supplementation frequency on cows consuming low-quality forage: Performance, grazing behavior, and variation in supplement intake. *Journal of Animal Science* 83, 1715–1725.

Schütz, K., Clark, K., Cox, N., Matthews, L., and Tucker, C. (2010) Responses to short-term exposure to simulated rain and wind by dairy cattle: Time budgets, shelter use, body temperature and feed intake. *Animal Welfare* 19, 375–383.

Selk, G., Wettemann, R., Lusby, K., Oltjen, J., Mobley, S., Rasby, R., and Garmendia, J.

(1988) Relationship among weight change, body condition and reproductive performance of range beef cows. *Journal of Animal Science* 66, 3153–3159.

Senft, R.L., Coughenour, M.B., Bailey, D.W., Rittenhouse, L.R., Sala, O.E., and Swift, D.M. (1987) Large herbivore foraging and ecological hierarchies. *BioScience* 37, 789–799.

Seo, S.N. (2008) Measuring impacts and adaptations to climate change: A structural Ricardian model of African livestock management. *Agricultural Economics* 38, 151–165.

Seo, S.N., McCarl, B.A., and Mendelsohn, R. (2010) From beef cattle to sheep under global warming? An analysis of adaptation by livestock species choice in South America. *Ecological Economics* 69, 2486–2494.

Siebert, B., and Macfarlane, W. (1975) Dehydration in desert cattle and camels. *Physiological Zoology* 48, 36–48.

Stephenson, M.B. (2014) Evaluation of alternative targeted cattle grazing practices and social association patterns of cattle in the western United States. PhD Dissertation, New Mexico State University, Las Cruces, NM.

Stuth, J., Jama, A., and Tolleson, D. (2003) Direct and indirect means of predicting forage quality through near infrared reflectance spectroscopy. *Field Crops Research* 84, 45–56.

Teague, W., Dowhower, S., Baker, S., Haile, N., DeLaune, P., and Conover, D. (2011) Grazing management impacts on vegetation, soil biota and soil chemical, physical and hydrological properties in tall grass prairie. *Agriculture, Ecosystems & Environment* 141, 310–322.

Thomas, J.W. (1979) Introduction. In: Thomas J.W. (Ed.) *Wildlife habitats in managed forests the Blue Mountains of Oregon and Washington*, Forest Service Handbook 553. USDA Forest Service, Washington, DC. pp. 10–21.

Titgemeyer, E., Drouillard, J., Greenwood, R., Ringler, J., Bindel, D., Hunter, R., and Nutsch, T. (2004) Effect of forage quality on digestion and performance responses of cattle to supplementation with cooked molasses blocks. *Journal of Animal Science* 82, 487–494.

Valentine, K.A. (1947) Distance from water as a factor in grazing capacity of rangeland. *Journal of Forestry* 45, 749–754.

Vallentine, J.F. (2001) *Grazing management*, 2nd edn. Academic Press, San Diego, CA.

Van Soest, P.J. (1994) *Nutritional ecology of the ruminant*, 2nd edn. Cornell University Press, Ithaca, NY.

VanWagoner, H.C., Bailey, D.W., Kress, D.D., Anderson, D.C., and Davis, K.C. (2006) Differences among beef sire breeds and relationships between terrain use and performance when daughters graze foothill rangelands as cows. *Applied Animal Behaviour Science* 97, 105–121.

Veissier, I., Butterworth, A., Bock, B., and Roe, E. (2008) European approaches to ensure good animal welfare. *Applied Animal Behaviour Science* 113, 279–297.

Villalba, J.J., and Provenza, F.D. (1999) Nutrient-specific preferences by lambs conditioned with intraruminal infusions of starch, casein, and water. *Journal of Animal Science* 77, 378–387.

Villalba, J.J., and Provenza, F.D. (2005) Foraging in chemically diverse environments: Energy, protein, and alternative foods influence ingestion of plant secondary metabolites by lambs. *Journal of Chemical Ecology* 31, 123–138.

Villalba, J.J., Provenza, F.D., and Bryant, J.P. (2002) Consequences of the interaction between nutrients and plant secondary metabolites on herbivore selectivity: Benefits or detriments for plants? *Oikos* 97, 282–292.

Walker, J.W. (2010) Primer on near infrared spectroscopy. Chapter 1. In: Walker, J. and Tolleson, D. (Eds.) *Shining light on manure improves livestock and land management,* Texas AgriLife Resarch Tech. Bull. SANG-2010-0250. Texas AgriLife Research, San Angelo, TX. pp. 1–7.

Wesley, R.L., Cibils, A.F., Pollak, E.R., Cox, S.H., Mulliniks, T., Petersen, M.K., and Fredrickson, E. (2008) Measures of daily distribution patterns of cow calf pairs using global positioning systems on both cows and calves. In: *Proceedings of the Joint Meeting of the Society for Range Management and the America Forage and Grassland Council.* Society for Range Management, Denver, CO.

White, N., Sutherst, R.W., Hall, N., and Whish-Wilson, P. (2003) The vulnerability of the Australian beef industry to impacts of the cattle tick (Boophilus microplus) under climate change. *Climatic Change* 61, 157–190.

Williams, R.E. (1954) Modern methods of getting uniform use of ranges. *Journal of Range Management* 7, 77–81.

Willms, W., Smoliak, S., and Dormaar, J. (1985) Effects of stocking rate on a rough fescue grassland vegetation. *Journal of Range Management* 38, 220–225.

Winchester, C., and Morris, M. (1956) Water intake rates of cattle. *Journal of Animal Science* 15, 722–740.

Winder, J., Walker, D., and Bailey, C. (1996) Effect of breed on botanical composition of cattle diets on Chihuahuan desert range. *Journal of Range Management* 49, 209–214.

Winder, J.A., Walker, D.A., and Bailey, C.C. (1995) Genetic aspects of diet selection in the Chihuahuan desert. *Journal of Range Management* 48, 549–553.

Young, B., and Degen, A. (1980) Ingestion of snow by cattle. *Journal of animal science* 51, 811–815.

chapter five

Impacts of toxic plants on the welfare of grazing livestock

James A. Pfister[1], Benedict T. Green[1], Kevin D. Welch[1], Frederick D. Provenza[2], and Daniel Cook[1]

INTRODUCTION

Interest in farm animal welfare has been increasing for several decades (Matthews, 1996). Animal health is an integral part of animal welfare (Stafford, 2014), but the concept of animal welfare has evolved from an emphasis on physical health, and coping ability (Broom, 1991) to a greater sensitivity to and recognition of animals' experiences of positive and negative feelings and emotions (Nicks and Vandenheede, 2014; Wemelsfelder and Mullan, 2014). The "five freedoms" provide an important framework for assessing animal welfare in livestock (Veissier et al., 2008), including: 1) freedom from thirst, hunger, and malnutrition; 2) freedom from discomfort; 3) freedom from pain, injury, and disease; 4) freedom to express normal behavior; and 5) freedom from fear and distress. One could argue that over-ingestion of plant toxins may negatively affect all of these freedoms. Some aspects (e.g., dietary selection) of the five freedoms may apply less rigorously

to extensively grazed animals, because animals freely grazing in extensive areas of pasture and rangeland can adapt behaviorally and physiologically as they interact with their environment, including toxic plants (Provenza, 2008; Ohl and Van der Staay, 2012).

The public perception is that the welfare of extensively grazed livestock is greatly enhanced because of their ability to express natural behaviors compared to more intensively managed animals (Matthews, 1996). However, relatively little work has been done to evaluate animal welfare within extensive systems (Fraser, 2008). Further, livestock grazing on rangelands may encounter conditions very different from either their early background or environment under which their species evolved, including encounters with plant secondary metabolites (PSMs) that occur in virtually all plants (Provenza et al., 2003). The ingestion of toxic plants by livestock grazing in extensive environments has the potential to adversely impact animal health through increased stress and suffering, as well as the

[1] USDA-ARS Poisonous Plant Research Laboratory, Logan, USA
[2] Utah State University, Wildland Resources Department, Logan, USA

potential to benefit animal welfare at low doses of PSMs (Provenza and Villalba, 2010; Durmic and Blache, 2012). Animals that can cope and adapt to encounters with toxic plants (Provenza et al., 2003) may not experience more than a temporary reduction in welfare and may benefit from mild stress imposed by secondary compounds (Provenza et al., 2015). However, animals are likely to experience short-term or long-term suffering if they are unable to cope with the challenges from encounters with toxic plants (Chahl and Kirk, 1975; Cromer and McIntyre, 2008; Roger, 2008). For example, introducing cattle breeds adapted to temperate climates into tropical or subtropical climates with subsequent exposure to unfamiliar plants and PSMs could result in compromised welfare and reduced production (Eisler et al., 2014). These issues may impact over one billion domestic livestock grazed on pasture and rangeland in the developed and developing world (Steinfeld et al., 2006), many of these animals in extensive systems.

Do livestock experience distress and pain (Chahl and Kirk, 1975; Cromer and McIntyre, 2008) upon ingesting toxic plants? Molony and Kent (1997) define animal pain as

> an aversive sensory and emotional experience representing an awareness by the animal of damage or threat to the integrity of its tissues; it changes the animal's physiology and behavior to reduce or avoid damage, to reduce the likelihood of recurrence and to promote recovery.

Livestock as prey species tend to be stoic when experiencing distress and pain (Weary et al., 2006), so it can be difficult to detect when they are experiencing noxious stimuli (FASS, 2010). Neural mechanisms of pleasure and reward are mediated by similar brain circuits in humans and other animals (Berridge and Kringelbach, 2008). Plant toxins may activate nerves consistent with causing pain (Chahl and Kirk, 1975; Cromer and McIntyre, 2008), and they apparently cause nausea in livestock (Provenza et al., 1994). Nonetheless, it is difficult to determine, let alone describe, how livestock experience the positive or aversive consequences of ingesting various toxic plants. Weary et al. (2006) outline three classes of pain response, including 1) pain-induced behavior such as vocalization; 2) a decline in frequency or magnitude of specific behaviors such as reduced movement; and 3) changes in choice or preference. Pain usually acts as a warning system for physiological distress, allowing animals to respond to and possibly escape from adverse stimuli. However, in the case of ingested toxins, there is little or no opportunity to flee; adjustments can only be made in subsequent choices related to ingestion of the plant (Provenza et al., 1992; Provenza, 1995a). In this review we assume that animals experience some unknown levels of stress or distress when ingestion of toxic plants exceeds a certain threshold and results in physiological responses such as tissue, nerve, or muscle damage, convulsions, GI distress, or if the intoxication results in endogenous production of proalgesic agents (Chahl and Kirk, 1975; Cromer and McIntyre, 2008; Basbaum et al., 2009). A reduction in animal welfare may be manifest through "pain-guarding" such as limping (Grandin and Deesing, 2003), or reductions in normal behaviors such as ruminating, grazing, suckling, exploring, or play behavior (Rushen, 1996). Further, abnormal behaviors such as depression, excessive vocalization, recumbency, ataxia, lack of motor activity, or excessive locomotor activity (Molony and Kent, 1997) may signal reduced well-being (Gonyou, 1994).

There is no doubt that animal welfare and assessments of welfare fall along a continuum from negative to positive (Ohl and Van der Staay, 2012). Assessments of welfare may be subjective (Fitzpatrick et al., 2006; Hewson, 2003a), or quantitative (Molony and Kent, 1997; Fitzpatrick et al., 2006; Stubsjøen et al., 2009). Animal scientists use various indicators of welfare, encompassing a wide range of possible problems from abnormal behavior, disease, production failure, and poor emotional states (Veissier and Miele, 2014). It seems clear that livestock producers and veterinarians, as the frontline in animal care, have a moral imperative to assess animal welfare (Bath 1998, Hewson 2003b), even though it may be easier to simply assess disease states (Hewson 2003a) using clinical signs from ingestion of toxic plants. In this chapter we review grazing animal welfare in relation to selected toxic plants, highlight areas where improvements in welfare can be made, and identify potential research needs.

NEUROLOGICAL EFFECTS

Three genera in the Fabaceae family, *Astragalus, Oxytropis*, and *Swainsona*, contain plant species that are toxic to grazing livestock because they contain the indolizidine alkaloid swainsonine (Molyneux and James, 1982). Swainsonine also has been reported in the Convolvulaceae and the Malvaceae families, including *Ipomoea, Turbina*, and *Sida* species (Cook et al., 2014). A fungal endophyte, *Undifilum* species has been shown to produce swainsonine in *Astragalus, Oxytropis*, and *Swainsona*, but different classes of endophytes produce swainsonine in other plant genera (Cook et al., 2014). Clinical signs and pathology of this neurologic disease are similar in animals intoxicated by any swainsonine-containing plant (often termed locoism; Stegelmeier et al., 1999). Swainsonine inhibits the cellular enzymes alpha-mannosidase and mannosidase II, thus altering glycoprotein processing and resulting in lysosomal storage disease (Dorling et al., 1980). Consumption of swainsonine-containing plants by grazing animals leads to chronic and debilitating neurologic disease characterized by weight loss, depression, altered behavior, decreased libido, infertility, abortion, and birth defects (James, 1972).

The effect of prolonged ingestion of even low doses of swainsonine on animal welfare is often profound (James, 1972; Pfister et al., 2003) and irreversible (Van Kampen and James, 1972), as continued consumption leads to reduced feed intake, severe behavioral depression, and eventual death. Abnormal behaviors are common such as increased startle response, intention tremors, reduced proprioception, extreme nervousness and agitation without reasonable cause, opisthotonos (star gazing), and misadventure from ataxia and malnutrition (James, 1972; Pfister et al., 1996). Pregnant animals that consume swainsonine-containing plants during gestation are adversely affected, and fetotoxicity is common (Panter et al., 1999). The affected dams often do not bond normally with their offspring at birth, and their neonates are typically weak and unable to nurse at birth (Pfister et al. 2006 a, b; Gotardo et al., 2011). Horses appear to be particularly susceptible to intoxication from swainsonine-containing plants, as they become behaviorally unstable and anorectic (Pfister et al., 2003). Information on the potentially detrimental effects of intoxication on human-equine interactions is lacking. Livestock poisoned by swainsonine-containing plants likely experience distress initially and progressively as the poisoning advances because of reduced motor control and proprioception.

Research on swainsonine-containing plants has been ongoing for many decades, but quantitative information to evaluate specific aspects of this toxin on animal welfare is lacking. For example, examination of intoxicated horses (e.g., impaired mobility via gait analysis) in concert with animals' selection of analgesic drugs in feeds could provide important information on pain levels (Wemelsfelder and Mullan, 2014). Further and more detailed work on ingestive behavior should be conducted (Pfister et al., 1996), including studies to determine if consumption of swainsonine-containing plants provides positive reinforcement (i.e., addiction; Huxtable and Dorling, 1982) using rodent models of self-administration (Quick et al., 2011).

The water hemlocks (*Cicuta* spp. and the related dropworts *Oenanthe* spp.) are likely the most violently toxic plants encountered by grazing livestock in North America and the UK (Davis et al., 2009). Water hemlock often grows in wet habitats like creeks, ditches, canals, or ponds (Panter et al., 2011). The major toxin is cicutoxin which acts in the CNS.

Pharmacologically, cicutoxin acts as a competitive γ-aminobutyric acid (GABA$_A$) receptor blocker. γ-aminobutyric acid is the main inhibitory neurotransmitter in the CNS so blockade of CNS inhibitory neural tone by cicutoxin results in excitotoxicity and other adverse sequela (Uwai et al., 2000), including overstimulation and seizures (Schep et al., 2009; Panter et al., 2011).

Clinical signs of water hemlock-induced poisoning include excessive salivation, nervousness, tremors, pilo-erection, frequent urination and defecation followed by incoordination, respiratory paralysis, muscular weakness, and convulsive seizures interspersed by intermittent periods of relaxation. In fatal intoxications, a final paralytic seizure results in anoxia and death (Panter et al., 1996; Gung and Omollo, 2009). Human reports indicate that mild toxicity from water hemlock produces nausea, abdominal pain, and epigastric distress within 15–90 minutes (Landers et al., 1985; Rizzi et al., 1991). In severe intoxications, profuse salivation, perspiration, bronchial secretion, and respiratory distress leading to cyanosis develop soon after ingestion (Heath, 2001). Interestingly, humans that are poisoned and recover report amnesia related to the seizures (Landers et al., 1985). No treatment can reverse the intoxication, although barbiturates (Panter et al., 1996) and diazepam (North and Nelson, 1985), both positive allosteric modulators of GABA$_A$ receptors (Sigel and Steinmann, 2012), may provide clinical relief and prevent fatal seizures. Assuming the pain and physical distress reported by humans is applicable to ruminant livestock, ingestion of water hemlock likely causes severe distress and pain to intoxicated animals. Anxiety that is not the result of direct pain may be severe in this type of intoxication because of the generalized loss of control experienced by animals undergoing grand-mal type seizures and muscular convulsions. Further, ictal pain related to seizures is not common but may occur in human patients (Bell et al., 1997).

MYOPATHIES

Ingestion of white snakeroot (*Ageratina altissima*) or rayless goldenrod (*Isocoma pluriflora*) equivalent to 5% of body weight over multiple days or consumption of milk products from lactating females eating these plants causes a condition in humans and livestock commonly called "trembles" (Lee et al., 2009; Davis et al., 2013). The putative toxins are likely benzofuran ketones, but the exact toxin(s) within this complex of compounds is not known (Lee et al., 2010). Clinically-affected animals show depression, inappetence, and muscle tremors especially after exercise or when stressed (Sharma et al., 1998). Intoxicated animals have an altered gait and posture, are excitable, and are reluctant to move, eventually becoming recumbent (Meyerholtz et al., 2011). Severe skeletal muscle necrosis causes the tremors, and the eventual death of severely affected animals results from cardiotoxicity. Horses (Finno et al., 2006) and goats (Stegelmeier et al., 2010) appear to be very sensitive to adverse effects when consuming these toxic plants.

"Milk sickness" that occurs when people drink the milk from animals consuming these plants has historically caused many deaths, including Abraham Lincoln's mother, in parts of the US where the plants are prevalent (Burrows and Tyrl, 2013). Humans have reported leg pain, cramps, vomition, inappetence, tremors, and transient or permanent muscle weakness after ingesting contaminated milk or milk products (Christensex, 1965). The clinical signs and muscular weakness that characterize the chronic and acute intoxication from white snakeroot and rayless goldenrod in livestock appear to have the potential to cause unrelieved stress and pain in affected animals, as manifested by the animals' behavioral changes (i.e., head pressing, hunched back, and stiff gait). The cardiac dysfunction (Smetzer et al., 1983) that also occurs seems likely to exacerbate stress in conjunction with the other clinical signs.

EFFECTS ON THE GASTROINTESTINAL TRACT

Solanum (nightshades) is a large and diverse genus which includes some species that provide important human foods (e.g., potato, tomato), whereas others are toxic (e.g., *S. elaeagnifolium*, *S. dulcamara*; Buck et al., 1960; Keeler et al., 1990). Many *Solanum* spp. contain glycoalkaloids with several types of aglycones that are typically less toxic than the glycoside (Burrows and Tyrl, 2013). Toxic *Solanum* species cause digestive disturbance (Bassett and Munro, 1986), neurological degeneration (Tokarnia et al., 2002; Verdes et al., 2006), calcinosis (Done, et al., 1976), and birth defects (Gaffield and Keeler, 1996). Prolonged consumption of *Solanum* spp. often results in neurological deficits in livestock. Clinical signs are periodic episodes of ataxia, head and thoracic limb extension, opisthotonos, nystagmus, and falling. Neurological deficits caused by *Solanum* spp. do not resolve, rendering affected animals permanently damaged (Verdes et al., 2006). Human consumption of potatoes, eggplant, tomatoes, and other edible plants that contain poly-hydroxy alkaloids (Molyneux et al., 2007) can cause gastric upset (Asano et al., 1997). Gastric disturbances in livestock are characterized by depression, vomition, diarrhea, abdominal pain, and possibly cardiac arrhythmias (Burrows and Tyrl, 2013). Both gastric disturbances and neurological disorders seem likely to cause distress in severely affected animals.

HEPATIC INJURY/ PHOTOSENSITIZATION

Pyrrolizidine alkaloids (PAs) have been identified in over 6,000 plants found throughout the world (Stegelmeier, 2011). Most PA-containing plants are considered unpalatable to livestock. Consequently most PA-containing plants are grazed when other forage is lacking, and they may also contaminate feeds and grains if harvested with other plants (Wiedenfeld, 2011; Stegelmeier, 2011). Pyrrolizidine alkaloids as ingested by livestock are not toxic, but are bio-activated in the liver to the toxic dehy-dropyrrolizidine alkaloids or pyrroles (Wiedenfeld, 2011). Chronic exposure to low doses of toxic PAs can lead to the accumulation of pyrrole–protein adducts and ultimately result in liver damage (Ruan et al., 2014). Intoxication by PAs can cause three different diseases in livestock: 1) acute intoxication with liver necrosis; 2) chronic liver disease with fibrosis causing photosensitization and ascites; and 3) chronic copper intoxication with hemolysis, jaundice, and hemoglobinuria (Anjos et al., 2010). Pyrrolizidine alkaloids are also potent lung toxins (Huxtable, 1990). Juvenile livestock are especially susceptible to PA-intoxication because of high rates of liver metabolism (Stegelmeier and Panter, 2012). Further, suckling offspring may be poisoned even if the mother is not affected (Small et al., 1993). There may be a substantial lag period from ingestion of the PA-containing plant, and the appearance of clinical signs, sometimes even years after exposure (Stegelmeier, 2011). Even with no outward clinical signs, many animals that ingest PA-containing plants continue to have liver damage, resulting in chronic liver inflammation, fibrosis, and cirrhosis (Wiedenfeld, 2011). Death is a frequent outcome if liver damage is sufficiently severe or if the resulting photosensitization (i.e., skin lesions) from impaired liver function is severe and untreated (Knight et al., 1984). Photosensitization of lightly pigmented body parts, or entire animals that are light skinned, frequently occurs with liver damage and subsequent circulation of phylloerythrin (Knight et al., 1984; Botha and Penrith, 2008). Compared to ruminants, horses are acutely sensitive to PA intoxication (Wiedenfeld, 2011); horses may also show neurological signs from PA toxicity (McLean, 1970). Even though goats and sheep are less sensitive to PA toxicity in comparison to cattle and horses (Stegelmeier, 2011), and can be used in bio-control grazing programs (Gardner et al., 2006), goats and sheep may be intoxicated by PA-containing plants (Anjos et al., 2010;

Maia et al., 2014). Interestingly, Anjos et al. (2010) reported that sheep develop resistance to PAs when given low PA doses for 20 days, although the resistance may only persist for 15 days post-dosing (Maia et al., 2014). Hepatotoxicity in humans from consumption of herbal preparations containing PAs (e.g., comfrey tea) is often characterized by epigastric pain with abdominal distension due to ascites (Stickel and Seitz, 2000). Livestock likely experience the same types of distress from advancing liver disease, and additional insult from skin lesions due to photosensitization, which may be exacerbated if photophobic animals are unable to stay out of the sun.

Saponins are found in many plants worldwide and exhibit a wide range of biological activities (Sparg et al., 2004). Saponin-induced hepatogenous photosensitization in livestock grazing on *Brachiaria* and *Panicum* spp. pastures is a serious problem in tropical and semi-tropical regions (Cheeke, 1995). Affected livestock develop lithogenic saponin crystals in the bile ducts and liver, leading to photophobia, photosensitization, and further liver and kidney damage (Wina et al., 2005). Saponins have been identified as the toxic agents affecting sheep that graze *Panicum* spp. grasses in the United States, New Zealand, Australia, Norway, South Africa, and South America (Wina et al., 2005). Outbreaks of poisoning and death in sheep, goats, buffalo, and cattle have been observed since the introduction of *Brachiaria* spp. grasses in Brazil (Driemeier et al., 2002; Riet-Correa et al., 2011). In Brazil, *Panicum* spp. (e.g., *P. dichotomiflorum*) and *Brachiaria* (primarily *B. decumbens* and *brizantha*) have been planted on over 60 million ha of pasture. Left untreated, the mortality

and morbidity from saponin-induced photosensitization likely causes substantial distress and pain, particularly in photophobic animals that graze in extensive tropical systems where they are exposed to direct sunlight.

MYCOTOXINS

Fungal endophytes associated with plants are often a source of secondary compounds in plants including potent toxins (Schardl et al., 2014). The magnitude of livestock losses attributed to fungal endophytes makes this the largest animal health-related production cost for grazing livestock worldwide (Poore and Washburn, 2013). The most economically important fungal-endophyte system is that of the genera *Epichloë* (external spore-producing fungi) and *Neotyphodium* (endophytic fungi that grow within a plant) with many cool season grasses (Poaceae, subfamily Poöideae) such as *Lolium* spp. (Bush et al., 1997). *Epichloë* and *Neotyphodium* species produce four classes of toxins in their symbiotic associations with plants: ergot alkaloids, indole-diterpenes, loline alkaloids, and peramine (Aly et al., 2010). Ergot alkaloids are also present in some plants in the morning glory family (Convolvulaceae) (Kucht et al., 2004). Ingestion of grasses contaminated with ergot peptide alkaloids causes three types of intoxication: a gangrenous syndrome affecting circulation in the extremities, a convulsive syndrome affecting the CNS, and a hyperthermic syndrome (summer slumps) affecting mainly cattle (Belser-Ehrlich et al., 2013). Gangrenous ergotism has a significant mortality rate, whereas hyperthermic ergotism is not often fatal. Ergot alkaloids act at serotonin receptors, and activating these receptors has been linked to hallucinations (Vollenweider et al., 1998), and nausea (Johnston et al., 2014). Ergot-type alkaloids probably cause nausea in affected livestock, because antiemetic drugs increase intake of infected forages (Aldrich et al., 1993). Cattle consuming contaminated grasses show numerous adverse effects including lower tolerance to ambient temperatures outside the animals' thermoneutral zone, and negative effects on forage intake, weight gain, and milk production (Strickland et al., 2011). When ingested by humans, ergot-type alkaloids produce intense burning sensations (known as St. Anthony's fire), abdominal pain, and death (Belser-Ehrlich et al., 2013). It seems likely that grazing animals would also feel distress and pain from similar symptoms. There are indications that some *Ipomoea* species including *Ipomoea muelleri* and *Ipomoea asarifolia* cause a tremorgenic syndrome in livestock (Tortelli et al., 2008; Carvalho de Lucena et al., 2014), and some reports suggest that these *Ipomoea* species contain indole-diterpene alkaloids (Schardl et al., 2013).

SUDDEN DEATH IN LIVESTOCK

Approximately 50 plant species worldwide, primarily in Africa, Australia, and South America, contain monofluoroacetate (MFA) and cause sudden death in livestock. Monofluoroacetate in plants is metabolized in the rumen to fluorocitrate which inhibits mitochondrial aconitase and blocks cellular respiration in the tricarboxylic acid cycle (Goncharov

et al., 2006). Clinical signs of livestock poisoned by MFA-containing plants appear in 4 to 24 hours, and include loss of balance, ataxia, tachycardia, and labored breathing (Goncharov et al., 2006). Intoxicated livestock often collapse after rising, sometimes remaining in lateral recumbence and pedaling in an attempt to stand up. Death usually occurs shortly thereafter. In less severe cases, a reluctance to walk or move, frequent posture changes, heart palpitations, and a visible pulsing jugular vein are also observed. No current therapy can reverse or prevent acute intoxication other than reducing ingestion of the MFA-containing plant (Lee et al., 2014). Reducing exposure to MFA-containing plants primarily involves changing animal management. Overgrazing and/ or drought are frequently implicated in sudden death by MFA-containing plant species (Lee et al., 2014). Some MFA-containing plants are very palatable (e.g., *Palicourea* in Brazil), and consumption is likely to increase if other forage is scarce because of drought, intensive grazing, or poor management. Animals poisoned by MFA-containing plants likely experience substantial distress initially and progressively as the poisoning advances because of a lack of motor control and inability to breathe normally. Marks et al. (2000) reported that apparent suffering in red foxes from fluoroacetate poisoning could be ameliorated by the anxiolytic/ sedative diazepam.

Larkspurs (*Delphinium* species) cause serious cattle losses in western North America (Pfister et al., 1999). Larkspurs typically contain various norditerpenoid alkaloids, and much of the toxicity is attributed to methyllycaconitine (MLA) (Welch et al., 2008). Receptor binding and in vitro functional studies have shown that nicotinic acetylcholine receptors (nAChR, particularly α7 receptors) in the muscle and brain are blocked by MLA and related alkaloids (Dobelis et al., 1999). Nicotinic AChR receptors are involved in many physiological functions including nociception (Badio and Daly, 1994). The primary result of larkspur toxicosis in livestock is neuromuscular paralysis. Clinical signs of intoxication in cattle include muscular weakness and trembling, straddled stance, periodic collapse into sternal recumbency, respiratory difficulty, progressing to lateral recumbency, and death. Death is hastened as poisoned animals become laterally recumbent, with accompanying posture-induced respiratory distress (e.g., bloat). Animal-to-animal variation is substantial in response to a non-fatal but toxic (i.e., able to bring on collapse) dose of larkspur alkaloids (Green et al., 2014). Ingestion of MLA in low doses by cattle may potentiate nAChRs in the CNS to facilitate consumption (Green et al., 2011). Further, Pfister et al. (1989) reported that low doses of larkspur enhanced nutritional parameters in cattle. However, ingestion of larkspurs by cattle in amounts sufficient to cause collapse and recumbency likely produces distress because of a lack of motor control as animals lose mobility and righting ability, particularly for extended periods of time typical in larkspur intoxication. The complex pharmacology of the different alkaloids acting on one or more nAChR subtypes precludes making definitive statements about pain in affected animals. Similar alkaloids (e.g., nicotine, Sahley and Berntson, 1979) and nAChR agonists can be potent analgesics (Gao et al., 2010).

BIRTH DEFECTS AND REPRODUCTION

Numerous toxic plants adversely affect either pregnant livestock or the developing fetus when eaten during pregnancy (Panter et al., 2013), but the most dramatic and noticeable are teratogenic effects. A number of plant toxins cross the placental barrier and impact the developing fetus. However, the developing embryo and fetus appear to be incapable of suffering and pain, as pain only occurs in newborn animals with the onset of breathing (Mellor and Diesch, 2006). Therefore the primary concern should be on maternal effects of plant toxins, and subsequent negative effects on the neonate after birth. Lupine species (*Lupinus* spp. such as *leucophyllus* and *caudatus*) that contain the quinolizidine alkaloid anagyrine cause multiple congenital contractures (MCC, so-called crooked calf disease) when eaten during days 40–100 of gestation by pregnant cattle (Panter et al., 2009; Lee et al., 2007). Anagyrine (and piperidine alkaloid-containing plants such as *Nicotiana glauca*) essentially anesthetize the fetus; the lack of fetal movement during this critical period results in severe deformities (Panter et al., 1990). In the crooked calf syndrome, calves are born with a variety of deformities such as arthrogryposis, scoliosis, kyphosis, torticollis, and cleft palate (Panter et al., 1997). Many of these calves survive at birth, but the prognosis is poor (Barrier et al., 2012) for virtually all deformed animals as they are not able to nurse or graze normally. Mobility is severely compromised from twisted limbs and spine. Calves with cleft palate may undergo additional stress from forage impaction in the open palate, and even death from aspiration of milk or rumen contents into the lungs resulting in pneumonia (Lee et al., 2008). Goats are also susceptible to various teratogenic alkaloids (Panter et al., 1990), with new-borns showing skeletal malformations and cleft palate. One of the most spectacular examples of plant-caused birth defects is cyclopia in lambs caused by ingestion by pregnant ewes of *Veratrum californicum* (Binns et al., 1963). Lambs are affected only if the plant is consumed on the 13th or 14th day of gestation (Welch et al., 2009) with affected lambs having numer-ous craniofacial malformations highlighted by a single median eye and a peculiar, elongated proboscis (Welch et al., 2009). The malformation is caused by the steroi-dal alkaloid cyclopamine (Keeler, 1969), which is also under investigation as a potential cancer therapy (Ma et al., 2013).

In addition to birth defects, prenatal exposure to plant toxins in utero may adversely affect both maternal and neonate behaviors (Pfister et al., 2006a, b) with subsequent increases in morbidity and mortality in the offspring (Dwyer, 2014). Neonates may experience pain more intensely than older animals, thus welfare at parturition and shortly thereaf-ter are of particular concern (Mellor and Stafford, 2004). Swainsonine-containing plants ingested during gestation have deleterious effects on both maternal and neonate behaviors (Pfister et al., 2006b; Gotardo et al., 2011) including maternal–infant bonding (Pfister et al., 2006a) and ability of the neonate to suckle shortly after birth (Pfister et al., 2006b).

TOXIC MINERALS

Ingestion of excess minerals from plants can cause chronic (e.g., selenium:

Raisbeck et al., 1993) or acute toxicity (Davis et al., 2012) in grazing livestock. Animal welfare may be negatively impacted in cases of chronic selenosis, particularly in hoof stock such as horses, as hoof lesions and lameness may be severe (Witte and Will, 1993), resulting in the need to euthanize the affected animal. Diagnosis of mineral status using hair analysis may be feasible in horses (Asano et al., 2002), and has proven successful for selenium, as Davis et al. (2014) used hair analysis to trace sele-nium exposure for up to three years post-exposure.

IMPROVING THE GENETIC "FIT" OF GRAZING ANIMALS TO COPE WITH TOXINS IN THEIR ENVIRONMENT

One solution to losses of grazing live-stock from toxic plants involves the compatibility of animals to their environ-ment. Genetic selection of livestock has the potential to enhance (e.g., dehorn-ing; Spurlock et al., 2014) or diminish (Oltenacu and Algers, 2005) animal welfare. Future global climate change may adversely affect animal welfare, whereas genetic selection may amelio-rate such reductions in welfare. It seems possible to select grazing livestock that are better adapted to future climate change (Hayes et al., 2009; Amer, 2012). Matching animal species, breeds, or genotypes to the proper environment may provide one avenue to reduce the negative effects of toxic plants on animal welfare. Campbell et al. (2010) selected lines of goats based on their consump-tion of juniper; goats that consumed large amounts of juniper possessed the ability to excrete monoterpenes more quickly. This trait has an estimated heritability of 11 to 13% (Waldron et al., 2009; Ellis et al., 2005). Other researchers identified a single nucleotide polymorphism in the dopamine receptor D2 gene that is associated with resistance to fescue toxicosis (Campbell et al., 2014). Research on *Delphinium* spp. poisoning identified substantial animal-to-animal variation in five breeds of cattle for resistance and susceptibility to larkspur (Green et al., 2014). Preliminary work suggests that each individual animal's genetic predisposition for larkspur poisoning can be predicted from its DNA sequence (B.T. Green, unpublished data). The ultimate goal of this research is to identify a gene marker that will allow producers to identify susceptible and resistant animals by submitting a saliva, blood, or hair sample to a genetic testing laboratory. Beef producers could then use that information for marker-assisted selection to improve the genetics of their herds for their specific rangeland conditions. There are currently commercially available genetic tests for coat color, production traits, and meat quality (Snelling et al., 2013; Kuehn et al., 2011). This same approach may prove valuable for other toxic plants such as lupine (*Lupinus* spp.). Animal welfare may be enhanced when livestock producers can use genetic tools to select the animals best suited to graze on pastures and rangelands with specific toxic plants. In the meantime, producers can select for locally adapted animals by observing livestock performance on pastures and rangelands and selecting individuals and cultures best adapted to their circumstances.

CLIMATE CHANGE, TOXIC PLANTS, AND ANIMAL WELFARE

Animal welfare is influenced greatly by environmental conditions, and predicted global climate change will likely influence the interaction of poisonous plants and grazing livestock. Climate change is expected to result in changes in weather patterns as well as elevated atmospheric CO_2. These changes may influence the size and density of plants, both of which may influence toxicity (e.g., *Delphinium nuttallianum*; Pfister and Cook, 2011). Likewise, elevated CO_2 increases biomass of *Lolium perenne* (perennial ryegrass) and *Datura stramonium* (jimsonweed) (Brosi et al. 2011; Ziska et al. 2005) and may influence PSMs in plants (Ahuja et al. 2010). If the composition or concentration of PSMs in an individual plant increases or decreases, the relative toxicity of the plant will change as well (Brosi et al., 2011; Ziska et al., 2005; Hunt et al., 2005). Elevated CO_2 increased scopolamine but did not alter atropine concentrations in jimsonweed (Ziska et al., 2005). Lastly, a complicating factor is that climate change may influence the palatability and nutritional content of plants. Elevated CO_2 concentrations reduce crude protein content in plants (Hunt et al., 2005). Ultimately, it is difficult to predict how changes in weather patterns and carbon dioxide will affect individual plant species or plant populations. The effects of climate change on plant chemistry and ecological interactions are highly context- and species-specific (Lindroth, 2010). These cascading effects will differ among plant species, PSMs, and plant populations, and will undoubtedly impact animal/toxic plant interactions (Dearing, 2013). In essence, plants and herbivores will continue to evolve with one another as they have for eons as climates changed.

MITIGATION OF EFFECTS OF TOXIC PLANTS ON GRAZING LIVESTOCK WELFARE

Positive welfare implies that the animal has the capability and freedom to adapt as necessary to both positive and potentially negative stimuli in their environment (Ohl and Van der Staay, 2012). Improving animal welfare by reducing consumption of lethal or debilitating quantities of poisonous plants by livestock will necessarily involve dietary choices (Manteca et al., 2008). Livestock grazing in extensive systems with toxic plants have a myriad of choices to make daily about where, what, and how long to graze within the context of the available resources (Provenza, 1995a, b; 1996). Choices often will be limited because of ongoing changes in the foraging milieu. In many episodes of toxic plant ingestion, grazing livestock may have little choice but to consume poisonous plants because of various management circumstances (Holechek, 2002). Drought, overgrazing, and poor grazing management have been linked with consumption of poisonous plants for over 200 years (Holechek, 2002). Even so, there are a number of palatable toxic plants that will be ingested by livestock even with an abundance of other forage available for grazing (e.g., *Palicourea* spp. in South America; *Delphinium* spp. in North America; *Astragalus* spp. in North America and China). For many toxic plants, poisoning episodes do not occur through repeated ingestion of low doses, but by over-ingestion of one or several toxic doses (e.g., *Delphinium*, Pfister et al., 1999; *Crotalaria*, Anjos et al., 2010). Ingestion of small quantities of acutely toxic plants (e.g., *Delphinium barbeyi*, Pfister et al., 1989) or chronically intoxicating plants (e.g., *Crotalaria*, Anjos et al., 2010) may actually enhance animal welfare and performance through an improved nutritional state.

"All substances are poisons; there is none which is not a poison. The right dose differentiates a poison from a remedy" (quote attributed to Paracelsus, 1493–1541). Numerous ethnopharmacological and biochemistry studies show that PSMs may have inestimable benefits for animal and human health (Huffman, 2003; Forbey et al., 2009; Molyneux et al., 2007; Roulette et al., 2014). Ingestion of poisonous plants as part of a diverse diet can increase disease resistance because animals can choose forages with antimicrobial (Wallace, 2004), antiparasitic (Athanasiadou and Kyriazakis, 2004), or immunity-enhancing properties (Singh et al., 2003; Villalba and Provenza, 2007; Provenza and Villalba, 2010). Villalba and Provenza (2007) have argued that

> food selection in herbivores can be interpreted as the constant quest for substances in the external environment that provide a homeostatic benefit to the internal environment .... Under this view, nutrients, PSMs, medicines are all 'substances' with the same final utility: improving the welfare of cells and cellular processes that enable life. Thus, self-medication – as ingestion of nutrients and avoidance of PSMs – is another dimension of homeostatic behaviour in animals. Ingesting

nutrients and medicines are means to the same end – stay well.

Livestock grazing on rangelands with toxic plants are faced with the tremendous challenge of balancing potential benefits (nutrients, medicine) with potential negative effects (disease, intoxication) of ingestion of PSMs (Estell, 2010). The trade-offs between benefit and harm are very complex, as toxin class, dose, and the physiological state of the animal all interact to influence the outcome. For example, multiple PSMs present in an animal's diet may interact in an additive or synergistic manner, through potentiation, or as antagonists (Villalba and Provenza, 2007; Copani et al., 2013), thus benefiting animal health and welfare, or to negatively affect animal health (Welch et al. 2014).

CONCLUSIONS

Animal welfare and well-being will be enhanced when grazing livestock are placed in circumstances in which they can make dietary selections that meet their needs for nutrients, and allow for regulation of toxin intake below a stressful or harmful threshold (Manteca et al., 2008; Rutter, 2010). Fostering adequate choices on rangelands with toxic plants will reduce risk of stress and intoxication, and allow grazing animals to better cope with toxins because some food combinations may ameliorate the effects of higher concentrations of toxins in the diet (Villalba et al., 2010). In some circumstances, livestock can be trained to avoid consumption of toxic plants using conditioned food aversions (Provenza, 1995a, b; Ralphs and Provenza, 1999), such that adverse effects on welfare can be eliminated. Further, efforts to match the type of livestock in extensive systems with the pasture features, including toxic plants, may have utility for improving animal welfare, pasture diversity and toxic plant populations, and outcomes from grazing management (Rook et al., 2004).

It is clear that the well-being and welfare of animals grazing in extensive systems will be enhanced as those personnel involved in their care train themselves in close observation of feeding behaviors of healthy livestock and in selecting for animals locally adapted to the environments where they live. Close observation and understanding of the normal behavioral repertoire of healthy animals will aid greatly in being able to detect early signs of ingestion of harmful levels of poisonous plants, and provide management opportunities to make changes in order to reduce pain and distress in animals before clinical signs are apparent. New technologies, such as "virtual fences" (Anderson, 2007; Umstatter, 2011) may eventually provide livestock producers the ability to remotely interact with grazing livestock in extensive systems, and specifically to interact with livestock in such a way as to direct animal movement away from toxic plant populations, or to take other actions to reduce episodes of intoxication. In the final analysis, efforts by livestock producers to augment and enhance diet selection and choice through range and/or animal management schemes will reduce livestock losses to toxic plants, improve grazing animal welfare and performance, and subsequently increase the profitability of livestock operations.

LOOKING AHEAD: AREAS FOR FUTURE RESEARCH

In order to understand the impacts of plant toxins on the welfare of grazing livestock, there are a number of research questions that need attention. Some of these questions will require continued improvements in technology in order to make substantial progress (e.g., Illumina BovineHD Genotyping BeadChip). For example, there is a critical need to determine the extent to which various grazing behaviors and patterns of diet selection are learned or are under genetic control (Bailey et al., 2015), and to initiate research that will further our understanding of epigenetic influences on behavior in grazing animals. This information will inform decisions about matching animals phenotypically and genetically to potential toxic plant scenarios for the benefit of both animals and grazing lands. In addition, there is a need to continue research into genetic markers (Green et al., 2014) that may be used to make important management decisions about grazing resistant or susceptible animals in a particular toxic plant environment (Scholtz et al., 2013).

There is a continuing need to understand the complex physiological and pathological effects of plant toxins on animal health, including the mechanism of action of plant toxins, in order to improve diagnostic and prognostic assessments, and to improve recognition of adverse or positive effects of plant toxins on animal welfare. There is a growing body of evidence on the beneficial (i.e., medicinal; anthelmintic) aspects of ingestion of plant toxins at low doses (Villalba and Provenza, 2007); in addition there is a need to continue research into possible health benefits of ingesting multiple plant toxins in low doses (i.e., complementarity).

There has been little work done on measuring the stress response of grazing animals consuming chronic doses of toxins in their diet. There appear to be opportunities to correlate behavioral and quantitative measures of welfare (Van Reenen et al., 2005; Comin et al., 2013) with acute or chronic stress, although this approach has not always been successful (Van Reenen et al., 2013; Probst et al., 2014). Stress and the resulting activation of the hypothalamic-pituitary-adrenal (HPA) axis (for reviews see Smith and Vale, 2006; Meyer and Novak, 2012) induces the release of corticotrophin releasing factor (CRF), which in turn activates the release of adrenocorticotropic hormone (ACTH) to stimulate the production of glucocorticoid hormones such as cortisol which have been associated with stress responses in livestock (Minton, 1994). Research in humans suggests that there is potential to relate the level of stress with glucocorticoid concentrations in blood, hair, or feces (Kirschbaum et al., 2009; Russell et al., 2012). Determination of the concentrations of glucocorticoids in hair, blood, or feces (Mormède et al., 2007; Comin et al., 2011; Russell et al., 2012; Comin et al., 2013) of grazing livestock may have utility in future toxic plant research to assess stress. Recent work in dairy cattle relating genetics (Peric et al., 2013), or clinical condition (Comin et al., 2013) to cortisol levels in hair indicates the potential utility of this approach for livestock consuming toxic plants. Tall fescue (*Lolium arundinaceum*) is often infected with fungal endophytes that produce ergot alkaloids; these compounds can severely impact cattle welfare during summer. Fescue researchers have studied the relationship between ergot alkaloid toxicity and potential serum markers of intoxication (e.g., prolactin; corticosteroids; Browning et al., 2000; Strickland et al., 2011), but these markers have not consistently been predictive of intoxication or animal stress (Aiken et al., 2011). Studies should continue to investigate blood concentrations of specific plant toxins, or perhaps the whole animal effect of plant toxins, in relation to concentrations of glucocorticoids as a measure of stress.

Animal distribution is often a major issue when grazing livestock ingest toxic plants. In this regard, there is a need to continue and expand studies into adaptive uses of emerging technologies that may improve animal distribution across the landscape (Anderson, 2007; Umstatter, 2011), with the objective of reducing or avoiding over-ingestion of toxic plants. Further, there is also a need for more research into potential effects of climate change on animal physiology, toxic plant populations, and toxin concentrations within plants.

REFERENCES

Ahuja, I., de Vos, R.C.H., Bones, A.M., and Hall, R.D. (2010) Plant molecular stress responses face climate change. *Trends in Plant Science* 15, 664–674.

Aiken, G., Klotz, J., Looper, M., Tabler, S., and Schrick, F. (2011) Disrupted hair follicle activity in cattle grazing endophyte-infected tall fescue in the summer insulates core body temperatures. *The Professional Animal Scientist* 27, 336–343.

Aldrich, C.G., Rhodes, M.T., Miner, J.L., Kerley, M.S., and Paterson, J.A. (1993) The effects of endophyte-infected tall fescue consumption and use of a dopamine antagonist on intake, digestibility, body temperature, and blood constituents in sheep. *Journal of Animal Science* 71, 158–163.

Aly, A.H., Debbab, A., Kjer, J., and Proksch, P. (2010) Fungal endophytes from higher plants: a prolific source of phytochemicals and other bioactive natural products. *Fungal Diversity*, 41, 1–16.

Amer, P.R. (2012) Turning science on robust cattle into improved genetic selection decisions. *Animal* 6, 551–556.

Anderson, D.M. (2007) Virtual fencing—past, present and future. *Rangeland Journal* 29, 65–78.

Anjos, B.L., Nobre, V.M.T., Dantas, A.F.M., Medeiros, R.M.T., Oliveira Neto, T.S., Molyneux, R.J., and Riet-Correa, F. (2010) Poisoning of sheep by seeds of *Crotalaria retusa*: Acquired resistance by continuous administration of low doses. *Toxicon* 55, 28–32.

Asano, N., Kato, A., Matsui, K., Watson, A.A., Nash, R.J., Molyneux, R.J., Hackett, L., Topping, J., and Winchester, B. (1997) Effects of calystegines isolated from edible fruits and vegetables on mammalian liver glycosidases. *Glycobiology*, 7:1085–1088.

Asano, R., Suzuki, K., Otsuka, T., Otsuka, M., and Sakurai, H. (2002) Concentrations of toxic metals and essential minerals in the mane hair of healthy racing horses and their relation to age. *Journal of Veterinary Medical Science* 64, 607–610.

Athanasiadou, S., and Kyriazakis, I. (2004) Plant secondary metabolites: Antiparasitic effects and their role in ruminant production systems. *Proceedings of the Nutrition Society* 63, 631–639.

Badio, B., and Daly, J.W. (1994) Epibatidine, a potent analgetic and nicotinic agonist. *Molecular Pharmacology* 45, 563–569.

Bailey, D.W., Lunt, S., Lipka, A., Thomas, M.G., Medrano, J.F., Cánovas, A., Rincon, G., Stephenson, M.B., and Jensen, D. (2015) Genetic influences on cattle grazing distribution: Association of genetic markers with terrain use in cattle. *Rangeland Ecology & Management* 68, 142–149. In press.

Barrier, A.C., Ruelle, E., Haskell, M.J., and Dwyer, C.M. (2012) Effect of a difficult calving on the vigour of the calf, the onset of maternal behaviour, and some behavioural indicators of pain in the dam. *Preventive Veterinary Medicine* 103, 248–256.

Basbaum, A.I., Bautista, D.M., Scherrer, G., and Julius, D. (2009) Cellular and molecular mechanisms of pain. *Cell* 139, 267–284.

Bassett, I., and Munro, D. (1986) The biology of Canadian weeds. 78. *Solanum carolinense* L. and *Solanum rostratum* Dunal. *Canadian Journal of Plant Science* 66, 977–991.

Bath, G. (1998) Management of pain in production animals. *Applied Animal Behaviour Science* 59, 147–156.

Bell, W.L., Walczak, T.S., Shin, C., and Radtke, R.A. (1997) Painful generalised clonic and tonic-clonic seizures with retained consciousness. *Journal of Neurology, Neurosurgery and Psychiatry* 63, 792–795.

Belser-Ehrlich, S., Harper, A., Hussey, J., and Hallock, R. (2013) Human and cattle ergotism since 1900-symptoms, outbreaks, and regulations. *Toxicology and Industrial Health* 29, 307–316.

Berridge, K.C., and Kringelbach, M.L. (2008) Affective neuroscience of pleasure: reward in humans and animals. *Psychopharmacology (Berl).* 199, 457–480.

Binns, W., James, L.F., Shupe, J.L., and Everett, G. (1963) A congenital cyclopian-type malformation in lambs induced by maternal ingestion of a range plant, *Veratrum californicum*. *American Journal of Veterinary Research* 24, 1164–1175.

Botha, C., and Penrith, M.L. (2008) Poisonous plants of veterinary and human importance in southern Africa. *Journal of Ethnopharmacology* 119, 549–558.

Broom, D.M. (1991) Animal welfare: Concepts and measurement. *Journal of Animal Science* 69, 4167–4175.

Brosi, G.B., McCulley, R.L., Bush, L.P., Nelson, J.A., Classen, A.T., and Norby, R.J. (2011) Effects of multiple climate change factors on the tall fescue–fungal endophyte symbiosis: Infection frequency and tissue chemistry. *New Phytologist* 189, 797–805.

Browning, R. Jr., Gissendanner, S.J., and Wakefield, T. Jr. (2000) Ergotamine alters plasma concentrations of glucagon, insulin, cortisol, and triiodothyronine in cows. *Journal of Animal Science* 78, 690–698.

Buck, W., Dollahite, J., and Alien, T. (1960) *Solanum elaeagnifolium*, silver-leafed nightshade, poisoning in livestock. *Journal of the American Veterinary Medical Association* 137, 348–351.

Burrows, G.E., and Tyrl, R.J. (2013) *Toxic plants of North America*, 2nd edn. Wiley-Blackwell, Ames, Iowa.

Bush, L.P., Wilkinson, H.H., and Schardl, C.L. (1997) Bioprotective alkaloids of grass-fungal endophyte symbioses. *Plant Physiology* 114, 1–7.

Campbell, E.J., Frost, R.A., Mosley, T.K., Mosley, J.C., Lupton, C.J., Taylor, C.A. Jr., Walker, J.W., Waldron, D.F., and Musser, J. (2010) Pharmacokinetic differences in exposure to camphor after intraruminal dosing in selectively bred lines of goats. *Journal of Animal Science* 88, 2620–2626.

Campbell, B.T., Kojima, C.J., Cooper, T.A., Bastin, B.C., Wojakiewicz, L., Kallenbach, R.L., Schrick, F.N., and Waller, J.C. (2014) A single nucleotide polymorphism in the dopamine receptor D2 gene may be informative for resistance to fescue toxicosis in Angus-based cattle. *Animal Biotechnology* 25, 1–12.

Carvalho de Lucena, K.F., Rodrigues, J.M.N., Campos, É.M., Dantas, A.F.M., Pfister, J.A., Cook, D., Medeiros, R.M.T., and Riet-Correa, F. (2014) Poisoning by *Ipomoea asarifolia* in lambs by the ingestion of milk from ewes that ingest the plant. *Toxicon* 92, 129–132.

Chahl, L. A., and Kirk, E. (1975) Toxins which produce pain. *Pain* 1, 3–49.

Cheeke, P.R. (1995) Endogenous toxins and mycotoxins in forage grasses and their effects on livestock. *Journal of Animal Science* 73, 909–918.

Christensex, W.I. (1965) Milk sickness: A review of the literature. *Economic Botany* 19, 293–300.

Comin, A., Prandi, A., Peric, T., Corazzin, M., Dovier, S., and Bovolenta, S. (2011) Hair cortisol levels in dairy cows from winter housing to summer highland grazing. *Livestock Science* 138, 69–73.

Comin, A., Peric, T., Corazzin, M., Veronesi, M., Meloni, T., Zufferli, V., Cornacchia, G., and Prandi, A. (2013) Hair cortisol as a marker of hypothalamic-pituitary-adrenal axis activation in Friesian dairy cows clinically or physiologically compromised. *Livestock Science* 152, 36–41.

Cook, D., Gardner, D.R., and Pfister, J.A. (2014) Swainsonine-containing plants and their relationship to endophytic fungi. *Journal of Agricultural and Food Chemistry* 62, 7326–7334.

Copani, G., Hall, J., Miller, J., Priolo, A., and Villalba, J. (2013) Plant secondary compounds as complementary resources: Are they always complementary? *Oecologia* 172, 1041–1049.

Cromer, B.A., and McIntyre, P. (2008) Painful toxins acting at TRPV1. *Toxicon* 51, 163–173.

Davis, T.Z., Lee, S., Ralphs, M., and Panter, K. (2009) Selected common poisonous plants of the United States' rangelands. *Rangelands* 31, 38–44.

Davis, T.Z., Stegelmeier, B.L., Panter, K.E., Cook, D., Gardner, D.R., and Hall, J.O. (2012) Toxicokinetics and pathology of plant-associated acute selenium toxicosis in steers. *Journal of Veterinary Diagnostic Investigation* 24, 319–327.

Davis, T.Z., Green, B.T., Stegelmeier, B.L., Lee, S.T., Welch, K.D., and Pfister, J.A. (2013) Physiological and serum biochemical changes associated with rayless goldenrod (*Isocoma pluriflora*) poisoning in goats. *Toxicon* 76, 247–254.

Davis, T.Z., Stegelmeier, B.L., and Hall, J.O. (2014) Analysis in horse hair as a means of evaluating selenium toxicoses and long-term exposures. *Journal of Agricultural and Food Chemistry* 62, 7393–7397.

Dearing, M.D. (2013) Temperature-dependent toxicity in mammals with implications for herbivores: A review. *Journal of Comparative Physiology B* 183, 43–50.

Dobelis, P., Madl, J.E., Pfister, J.A., Manners, G.D., and Walrond, J.P. (1999) Effects of *Delphinium* alkaloids on neuromuscular transmission. *Journal of Pharmacology and Experimental Therapeutics* 291, 538–546.

Done, S., Döbereiner, J., and Tokarnia, C. (1976) Systemic connective tissue calcification in cattle poisoned by *Solanum malacoxylon*: A histological study. *The British Veterinary Journal* 132, 28–38.

Dorling, P.R., Huxtable, C.R., and Colegate, S.M. (1980) Inhibition of lysosomal a-mannosidase by swainsonine, an indolizidine alkaloid isolated from *Swainsona canescens*. *Biochemical Journal* 191, 649–651.

Driemeier, D., Colodel, E.M., Seitz, A.L., Barros, S.S., and Cruz, C.E. (2002) Study of experimentally induced lesions in sheep by grazing *Brachiaria decumbens*. *Toxicon* 40, 1027–1031.

Durmic, Z., and Blache, D. (2012) Bioactive plants and plant products: Effects on animal function, health and welfare. *Animal Feed Science and Technology* 176, 150–162.

Dwyer, C. (2014) Maternal behaviour and lamb survival: From neuroendocrinology to practical application. *Animal* 8, 102–112.

Eisler, M.C., Lee, M.Tarlton, J.F., Martin, G.B., Beddington, J., Dungait, J., Greathead, H., Liu, J., Mathew, S., and Miller, H. (2014) Agriculture: Steps to sustainable livestock. *Nature* 507, 32–34.

Ellis, C.R., Jones, R.E., Scott, C.B., Taylor, C.A., Walker, J.W., and Waldron, D.E. (2005) Sire influence on juniper consumption by goats. *Rangeland Ecology and Management* 58, 324–328.

Estell, R.E. (2010) Coping with shrub secondary metabolites by ruminants. *Small Ruminant Research* 94, 1–9.

Federation of Animal Science Societies (FASS). (2010) *Guide for the care and use of agricultural animals in research and teaching*, 3rd edn. Champaign, Illinois.

Finno, C.J., Valberg, S.J., Wünschmann, A., and Murphy, M.J. (2006) Seasonal pasture myopathy in horses in the midwestern United States: 14 cases (1998–2005). *Journal of the American Veterinary Medical Association* 229, 1134–1141.

Fitzpatrick, J., Scott, M., and Nolan, A. (2006) Assessment of pain and welfare in sheep. *Small Ruminant Research* 62: 55–61.

Forbey, J.S., Harvey, A.L., Huffman, M.A., Provenza, F.D., Sullivan, R., and Tasdemir, D. (2009) Exploitation of secondary metabolites by animals: A response to homeostatic challenges. *Integrative and Comparative Biology* 49, 314–328.

Fraser, D. (2008) Toward a global perspective on farm animal welfare. *Applied Animal Behaviour Science* 113, 330–339.

Gaffield, W., and Keeler, R.F. (1996) Induction of terata in hamsters by solanidane alkaloids derived from *Solanum tuberosum*. *Chemical Research in Toxicology* 9, 426–433.

Gao, B., Hierl, M., Clarkin, K., Juan, T., Nguyen, H., Valk, M.v.d., Deng, H., Guo, W., Lehto, S.G. and Matson, D. (2010). Pharmacological effects of nonselective and subtype-selective nicotinic acetylcholine receptor agonists in animal models of persistent pain. *Pain* 149, 33–49.

Gardner, D.R., Thorne, M.S., Molyneux, R.J., Pfister, J.A., and Seawright, A.A. (2006) Pyrrolizidine alkaloids in *Senecio madagascariensis* from Australia and Hawaii and assessment of possible livestock poisoning. *Biochemical Systematics and Ecology* 34, 736–744.

Goncharov, N.V., Jenkins, R.O., and Radilov, A.S. (2006) Toxicology of fluoroacetate: A review, with possible directions for therapy research. *Journal of Applied Toxicology* 26, 148–161.

Gonyou, H.W. (1994) Why the study of animal behaviour is associated with the animal welfare issue. *Journal of Animal Science* 72, 2171–2177.

Gotardo, A.T., Pfister, J.A., Ferreira, M.B., and Gorniak, S.L. (2011) Effects of prepartum ingestion of *Ipomoea carnea* on postpartum maternal and neonate behavior in goats. *Birth Defects Research Part B: Developmental and Reproductive Toxicology* 92, 131–138.

Grandin, T., and Deesing, M. (2003) Distress in animals: Is it fear, pain, or physical stress? Available at http://www.grandin.com/welfare/fear.pain.stress.html (accessed 4 November, 2014).

Green, B.T., Welch, K.D., Cook, D., and Gardner, D.R. (2011) Potentiation of the actions of acetylcholine, epibatidine, and nicotine by methyllycaconitine at fetal muscle-type nicotinic acetylcholine receptors. *European Journal of Pharmacology* 662, 15–21.

Green, B.T., Welch, K.D., Pfister, J.A., Chitko-McKown, C.G., Gardner, D.R., and Panter, K.E. (2014) Mitigation of larkspur poisoning on rangelands through the selection of cattle. *Rangelands* 36, 10–15.

Gung, B.W., and Omollo, A.O. (2009) A concise synthesis of R-(–)-cicutoxin, a natural 17-carbon polyenyne. *European Journal of Organic Chemistry* 8, 1136–1138.

Hayes, B.J., Bowman, P.J., Chamberlain, A.J., Savin, K., van Tassell, C.P., Sonstegard, T,S., and Goddard, M.E. (2009) A validated genome wide association study to breed cattle adapted to an environment altered by climate change. *PLoS One* 4, e6676.

Heath, K. (2001) A fatal case of apparent water hemlock poisoning. *Veterinary and Human Toxicology* 43, 35–36.

Hewson, C.J. (2003a) Can we assess welfare? *The Canadian Veterinary Journal* 44, 749–753.

Hewson, C.J. (2003b). How might veterinarians do more for animal welfare? *The Canadian Veterinary Journal* 44, 1000–1004.

Holechek, J.L. (2002) Do most livestock losses to poisonous plants result from "poor" range management? *Journal of Range Management* 55, 270–276.

Huffman, M.A. (2003) Animal self-medication and ethno-medicine: Exploration and exploitation of the medicinal properties of plants. *Proceedings of the Nutrition Society* 62, 371–381.

Hunt, M.G., Rasmussen, S., Newton, P.C., Parsons, A.J., and Newman, J.A. (2005) Near-term impacts of elevated CO_2, nitrogen and fungal endophyte-infection on *Lolium perenne* L. growth, chemical composition and alkaloid production. *Plant, Cell and Environment* 28, 1345–1354.

Huxtable, R.J. (1990) Activation and pulmonary toxicity of pyrrolizidine alkaloids. *Pharmacology and Therapeutics* 47, 371–389.

Huxtable, C., and Dorling, P. (1982) Poisoning of livestock by *Swainsona* spp.: Current status. *Australian Veterinary Journal* 59, 50–53.

James, L.F. (1972) Syndromes of locoweed poisoning in livestock. *Clinical Toxicology* 5, 567–573.

Johnston, K.D., Lu, Z., and Rudd, J.A. (2014) Looking beyond 5-HT$_3$ receptors: A review of the wider role of serotonin in the pharmacology of nausea and vomiting. *European Journal of Pharmacology* 722, 13–25.

Keeler, R. (1969) Teratogenic compounds of *Veratrum californicum*-VI: The structure of cyclopamine. *Phytochemistry* 8, 223–225.

Keeler, R.F., Baker, D.C., and Gaffield, W. (1990) Spirosolane-containing *Solanum* species and induction of congenital craniofacial malformations. *Toxicon* 28, 873–884.

Kirschbaum, C., Tietze, A., Skoluda, N., and Dettenborn, L. (2009) Hair as a retrospective calendar of cortisol production—increased cortisol incorporation into hair in the third trimester of pregnancy. *Psychoneuroendocrinology* 34, 32–37.

Knight, A.P., Kimberling, C.V., Stermitz, F.R., and Roby, M.R. (1984) *Cynoglossum officinale* (hound's-tongue)–a cause of pyrrolizidine alkaloid poisoning in horses. *Journal of the American Veterinary Medical Association* 185, 647–650.

Kucht, S., Groß, J., Hussein, Y., Grothe, T., Keller, U., Basar, S., K€onig, W.A., Steiner, U., and Leistner, E. (2004) Elimination of ergoline alkaloids following treatment of *Ipomoea asarifolia* (Convolvulaceae) with fungicides. *Planta* 219, 619–625.

Kuehn, L.A., Keele, J.W., Bennett, G.L., McDaneld, T.G., Smith, T.P., Snelling, W.M., Sonstegard, T.S., and Thallman, R.M. (2011) Predicting breed composition using breed frequencies of 50,000 markers from the US Meat Animal Research Center 2,000 Bull Project. *Journal of Animal Science* 89, 1742–50.

Landers, D., Seppi, K., and Blauer W. (1985) Seizures and death on a white river float trip: Report of water hemlock poisoning. *Western Journal of Medicine* 142, 637.

Lee, S.T., Cook, D., Panter, K.E., Gardner, D.R., Ralphs, M.H., Motteram, E.S., Pfister, J.A., and Gay, C.C. (2007) Lupine-induced "crooked calf disease" in Washington and Oregon: Identification of the alkaloid profiles in *Lupinus sulfureus*, *Lupinus leucophyllus*, and *Lupinus sericeus*. *Journal of Agricultural and Food Chemistry* 55, 10649–10655.

Lee, S.T., Panter, K.E., Gay, C.C., Pfister, J.A., Ralphs, M.H., Gardner, D.R., Stegelmeier, B.L., Motteram, E.S., Cook, D., and Welch, K.D. (2008) Lupine-induced crooked calf disease: The last 20 years. *Rangelands* 30, 13–18.

Lee, S.T., Davis, T.Z., Gardner, D.R., Stegelmeier, B.L., and Evans, T.J. (2009) Quantitative method for the measurement of three benzofuran ketones in rayless goldenrod (Isocoma *pluriflora*) and white snakeroot (*Ageratina altissima*) by high-performance liquid chromatography (HPLC). *Journal of Agricultural and Food Chemistry* 57, 5639–5643.

Lee, S.T., Davis, T.Z., Gardner, D.R., Colegate, S.M., Cook, D., Green, B.T., Meyerholtz, K.A., Wilson, C.R., Stegelmeier, B.L., and Evans, T.J. (2010) Tremetone and structurally related compounds in white snakeroot (*Ageratina altissima*): A plant associated with trembles and milk sickness. *Journal of Agricultural and Food Chemistry* 58, 8560–8565.

Lee, S.T., Cook, D., Pfister, J.A., Allen, J., Colegate, S.M., Riet-Correa, F., and Taylor, C.M. (2014) Monofluoroacetate-containing plants that are potentially toxic to livestock. *Journal of Agricultural and Food Chemistry* 62, 7345–7354.

Lindroth, R.L. (2010) Impacts of elevated atmospheric CO_2 and O_3 on forests: Phytochemistry, trophic interactions, and ecosystem dynamics. *Journal of Chemical Ecology* 36, 2–21.

Ma, H., Li, H.Q., and Zhang, X. (2013) Cyclopamine, a naturally occurring alkaloid, and its analogues may find wide applications in cancer therapy. *Current Topics in Medicinal Chemistry* 13, 2208–2215.

Maia, L.A., Pessoa, C.R.d.M., Rodrigues, A.F., Colegate, S., Dantas, A.F.M., Medeiros, R.M.T., and Riet-Correa, F. (2014) Duration of an induced resistance of sheep to acute poisoning by *Crotalaria retusa* seeds. *Ciência Rural* 44, 1054–1059.

Manteca, X., Villalba, J.J., Atwood, S.B., Dziba, L., and Provenza, F.D. (2008) Is dietary choice important to animal welfare? *Journal of Veterinary Behavior: Clinical Applications and Research* 3, 229–239.

Marks, C.A., Hackman, C., Busana, F., and Gigliotti, F. (2000) Assuring that 1080 toxicosis in the red fox (*Vulpes vulpes*) is humane: Fluoroacetic acid (1080) and drug combinations. *Wildlife Research* 27, 483–494.

Matthews, L. (1996) Animal welfare and sustainability of production under extensive conditions: A non-EU perspective. *Applied Animal Behaviour Science* 49, 41–46.

McLean, E.K. (1970) The toxic actions of pyrrolizidine (*Senecio*) alkaloids. *Pharmacological Reviews* 22, 429–483.

Mellor, D.J., and Diesch, T.J. (2006) Onset of sentience: The potential for suffering in fetal and newborn farm animals. *Applied Animal Behaviour Science* 100, 48–57.

Mellor, D., and Stafford, K. (2004) Animal welfare implications of neonatal mortality and morbidity in farm animals. *The Veterinary Journal* 168, 118–133.

Meyer, J.S., and Novak, M.A. (2012) Minireview: Hair cortisol: A novel biomarker of hypothalamic-pituitary-adrenocortical activity. *Endocrinology* 153, 4120–4127.

Meyerholtz, K.A., Burcham, G.N., Miller, M.A., Wilson, C.R., Hooser, S.B., and Lee, S.T. (2011) Development of a gas chromatography–mass spectrometry technique to diagnose white snakeroot (*Ageratina altissima*) poisoning in a cow. *Journal of Veterinary Diagnostic Investigation* 23, 775–779.

Minton, J.E. (1994) Function of the hypothalamic-pituitary-adrenal axis and the sympathetic nervous system in models of acute stress in domestic farm animals. *Journal of Animal Science* 72, 1891–1898.

Molony, V., and Kent, J. (1997) Assessment of acute pain in farm animals using behavioral and physiological measurements. *Journal of Animal Science* 75, 266–272.

Molyneux, R.J., and James, L. F. (1982) Loco intoxication: Indolizidine alkaloids of spotted locoweed (*Astragalus lentiginosus*). *Science* 216, 190–191.

Molyneux, R.J., Lee, S.T., Gardner, D.R., Panter, K.E., and James, L.F. (2007) Phytochemicals: The good, the bad and the ugly? *Phytochemistry* 68, 2973–2985.

Mormède, P., Andanson, S., Aupérin, B., Beerda, B., Guémené, D., Malmkvist, J., Manteca, X., Manteuffel, G., Prunet, P., van Reenen, C.G., Richard, S., and Veissier, I (2007) Exploration of the hypothalamic-pituitary-adrenal function as a tool to evaluate animal welfare. *Physiology and Behavior* 92, 317–339

Nicks, B., and Vandenheede, M. (2014) Animal health and welfare: Equivalent or complementary? *Revue Scientifique et Technique (International Office of Epizootics)* 33, 97–101.

North, D., and Nelson, R. (1985) Anticholinergic agents in cicutoxin poisoning. *Western Journal of Medicine* 143, 250.

Ohl, F., and Van der Staay, F.J. (2012) Animal welfare: At the interface between science and society. *The Veterinary Journal* 192, 13–19.

Oltenacu, P.A., and Algers, B. (2005) Selection for increased production and the welfare of dairy cows: Are new breeding goals needed? *AMBIO: A Journal of the Human Environment* 34, 311–315.

Panter, K., Keeler, R., Bunch, T., and Callan, R. (1990) Congenital skeletal malformations and cleft palate induced in goats by ingestion of *Lupinus, Conium* and *Nicotiana* species. *Toxicon* 28, 1377–1385.

Panter, K.E., Baker, D.C., and Kechele P.O. (1996) Water hemlock (*Cicuta douglasii*) toxicoses in sheep: Pathologic description and prevention of lesions and death. *Journal of Veterinary Diagnostic Investigation* 8, 474–480.

Panter, K.E., Gardner, D.R., Gay, C.C., James, L.F., Mills, R., Gay, J.M., and Baldwin, T.J. (1997) Observations of *Lupinus sulphureus*-induced "crooked calf disease" *Journal of Range Management* 50, 587–592.

Panter, K., James, L., Stegelmeier, B., Ralphs, M., and Pfister, J. (1999) Locoweeds: Effects on reproduction in livestock. *Journal of Natural Toxins* 8, 53–62.

Panter, K., Motteram, E., Cook, D., Lee, S., Ralphs, M., Platt, T., and Gay, C. (2009) Crooked calf syndrome: Managing lupines on rangelands of the Channel Scablands of east-central Washington State. *Rangelands* 31, 10–15.

Panter, K.E., Gardner, D.R., Stegelmeier, B.L., Welch, K.D., and Holstege, D.

(2011) Water hemlock poisoning in cattle: Ingestion of immature *Cicuta maculata* seed as the probable cause. *Toxicon* 57, 157–161.

Panter, K.E., Welch, K.D., Gardner, D.R., and Green, B.T. (2013) Poisonous plants: Effects on embryo and fetal development. *Birth Defects Research Part C: Embryo Today: Reviews* 99, 223–234.

Peric, T., Comin, A., Corazzin, M., Montillo, M., Cappa, A., Campanile, G., and Prandi, A. (2013) Short communication: Hair cortisol concentrations in Holstein-Friesian and crossbreed F1 heifers. *Journal of Dairy Science* 96, 3023–3027.

Pfister, J., and Cook, D. (2011) Influence of weather on low larkspur (*Delphinium nuttallianum*) density. *Journal of Agricultural Science* 3, 36–44.

Pfister, J., Adams, D., Arambel, M., Olsen, J., and James, L. (1989) Sublethal levels of toxic larkspur: Effects on intake and rumen dynamics in cattle. *Nutrition Reports International* 40, 629–636.

Pfister, J.A., Stegelmeier, B.L., Cheney, C.D., James, L.F., and Molyneux, R.J. (1996) Operant analysis of chronic locoweed intoxication in sheep. *Journal of Animal Science* 74, 2622–2632.

Pfister, J.A., Gardner, D.R., Panter, K.E., Manners, G.D., Ralphs, M.H., Stegelmeier, B.L., and Schoch, T.K. (1999) Larkspur (*Delphinium* spp.) poisoning in livestock. *Journal of Natural Toxins* 8, 81–94.

Pfister, J., Stegelmeier, B., Gardner, D., and James, L. (2003) Grazing of spotted locoweed (*Astragalus lentiginosus*) by cattle and horses in Arizona. *Journal of Animal Science* 81, 2285–2293.

Pfister, J., Astorga, J., Panter, K., Stegelmeier, B., and Molyneux, R. (2006a) Maternal ingestion of locoweed: I. Effects on ewe–lamb bonding and behaviour. *Small Ruminant Research* 65, 51–63.

Pfister, J., Davidson, T., Panter, K., Cheney, C., and Molyneux, R. (2006b) Maternal ingestion of locoweed: III. Effects on lamb behaviour at birth. *Small Ruminant Research* 65, 70–78.

Poore, M., and Washburn, S. (2013) Forages and pastures symposium: Impact of fungal endophytes on pasture and environmental sustainability. *Journal of Animal Science* 91, 2367–2368.

Probst, J.K., Spengler Neff, A., Hillmann, E., Kreuzer, M., Koch-Mathis, M., and Leiber, F. (2014) Relationship between stress-related exsanguination blood variables, vocalisation, and stressors imposed on cattle between lairage and stunning box under conventional abattoir conditions. *Livestock Science* 164, 154–158.

Provenza, F.D. (1995a) Postingestive feedback as an elementary determinant of food preference and intake in ruminants. *Journal of Range Management* 48, 2–17.

Provenza, F.D. (1995b) Tracking variable environments: There is more than one kind of memory. *Journal of Chemical Ecology* 21, 911–923.

Provenza, F.D. (1996) Acquired aversions as the basis for varied diets of ruminants foraging on rangelands. *Journal of Animal Science* 74, 2010–2020.

Provenza, F.D. (2008) What does it mean to be locally adapted and who cares anyway? *Journal of Animal Science* 86, E271–E284.

Provenza, F.D., and Villalba, J.J. (2010) The role of natural plant products in modulating the immune system: An adaptable approach for combating disease in grazing animals. *Small Ruminant Research* 89, 131–139.

Provenza, F.D., Pfister, J.A., and Cheney, C.D. (1992) Mechanisms of learning in diet selection with reference to phytotoxicosis in herbivores. *Journal of Range Management* 45, 36–45.

Provenza, F.D., Ortega-Reyes, L., Scott, C.B., Lynch J.J., and Burritt E.A. (1994) Antiemetic drugs attenuate food aversions in sheep. *Journal of Animal Science* 72, 1989–1994.

Provenza, F.D., Villalba, J.J., Dziba, L.E., Atwood, S.B., and Banner, R.E. (2003) Linking herbivore experience, varied diets, and plant biochemical diversity. *Small Ruminant Research* 49, 257–274.

Provenza, F.D., Gregorini, P., and Carvalho, P.C.F. (2015) Synthesis: Foraging decisions link the cells of plants, herbivores, and human beings. *Animal Production Science* 54. In press.

Quick, S.L., Pyszczynski, A.D., Colston, K.A., and Shahan, T.A. (2011) Loss of alternative non-drug reinforcement induces relapse of cocaine-seeking in rats: Role of dopamine D1 receptors. *Neuropsychopharmacology* 36: 1015–1020.

Raisbeck, M., Dahl, E., Sanchez, D., Belden, E., and O'Toole, D. (1993) Naturally occurring selenosis in Wyoming. *Journal of Veterinary Diagnostic Investigation* 5, 84–87.

Ralphs, M.H., and Provenza, F.D. (1999) Conditioned food aversions: Principles and practices, with special reference to social facilitation. *Proceedings of the Nutrition Society* 58, 813–820.

Riet-Correa, B., Castro, M.B., Lemos, R., Riet-Correa, G., Mustafa, V., and Riet-Correa, F. (2011) *Brachiaria* spp. poisoning of ruminants in Brazil. *Pesquisa Veterinária Brasileira* 31, 183–192.

Rizzi, D., Basile, C., Di Maggio, A., Sebastio, A., Introna, F., Rizzi, R., Scatizzi, A., De Marco, S., and Smialek, J.E. (1991) Clinical spectrum of accidental hemlock poisoning: Neurotoxic manifestations, rhabdomyolysis and acute tubular necrosis. *Nephrology Dialysis Transplantation* 6, 939–943.

Roger, P. (2008) The impact of disease and disease prevention on sheep welfare. *Small Ruminant Research* 76, 104–111.

Rook, A.J., Dumont, B., Isselstein, J., Osoro, K., WallisDeVries, M.F., Parente, G., and Mills, J. (2004) Matching type of livestock to desired biodiversity outcomes in pastures – a review. *Biological Conservation* 119, 137–150.

Roulette, C.J., Mann, H., Kemp, B.M., Remiker, M., Roulette, J.W., Hewlett,

B.S., Kazanji, M., Breurec, S., Monchy, D., and Sullivan, R.J. (2014) Tobacco use vs. helminths in Congo basin hunter-gatherers: Self-medication in humans? *Evolution and Human Behavior* 35, 397–409.

Ruan, J., Yang, M., Fu, P., Ye, Y., and Lin, G. (2014) Metabolic activation of pyrrolizidine alkaloids: Insights into the structural and enzymatic basis. *Chemical Research in Toxicology* 27, 1030–1039.

Rushen, J. (1996) Using aversion learning techniques to assess the mental state, suffering and welfare of farm animals. *Journal of Animal Science* 74, 1990–1995.

Russell, E., Koren, G., Rieder, M., Van Uum, S. (2012) Hair cortisol as a biological marker of chronic stress: Current status, future directions and unanswered questions. *Psychoneuroendocrinology* 37, 589–601.

Rutter, S.M. (2010) Review: Grazing preferences in sheep and cattle: Implications for production, the environment and animal welfare. *Canadian Journal of Animal Science* 90, 285–293.

Sahley, T.L., and Berntson, G.G. (1979) Antinociceptive effects of central and systemic administrations of nicotine in the rat. *Psychopharmacology* 65, 279–283.

Schardl, C.L., Young, C.A., Hesse, U., Amyotte, S.G., Andreeva, K., Calie, P.J., Fleetwood, D.J., Haws, D.C., Moore, N., and Oeser, B. (2013) Plant-symbiotic fungi as chemical engineers: Multi-genome analysis of the Clavicipitaceae reveals dynamics of alkaloid loci. *PLoS Genetics* 9, e1003323.

Schardl, C.L., Chen, L., and Young, C. A. (2014) Fungal endophytes of grasses and morning glories, and their bioprotective alkaloids. In: Osbourn, A., Goss, R.J., and Carter, G.T. (Eds.) *Natural products: Discourse, diversity, and design.* John Wiley and Sons, Inc., Ames, Iowa. pp. 125–145.

Schep, L.J., Slaughter, R.J., Becket, G., and Beasley, D.M.G. (2009) Poisoning due to water hemlock. *Clinical Toxicology* 47, 270–278.

Scholtz, M., Maiwashe, A., Neser, F., Theunissen, A., Olivier, W., Mokolobate, M., and Hendriks, J. (2013) Livestock breeding for sustainability to mitigate global warming, with the emphasis on developing countries. *South African Journal of Animal Science* 43, 269–281.

Sharma, O.P., Dawra, R.K., Kurade, N.P., and Sharma, P.D. (1998) A review of the toxicosis and biological properties of the genus *Eupatorium. Natural Toxins* 6, 1–14.

Sigel, E., and Steinmann, M.E. (2012) Structure, function, and modulation of GABA$_{(A)}$ receptors. *Journal of Biological Chemistry* 287, 40224–40231.

Singh, B., Bhat, T.K., and Singh, B. (2003) Potential therapeutic applications of some antinutritional plant secondary metabolites. *Journal of Agricultural and Food Chemistry* 51, 5579–5597.

Small, A., Kelly, W., Seawright, A., Mattocks, A., and Jukes, R. (1993) Pyrrolizidine alkaloidosis in a two month old foal. *Journal of Veterinary Medicine Series A* 40, 213–218.

Smetzer, D.L., Coppock, R.W., Ely, R.W., Duckett, W.M., and Buck, W.B. (1983) Cardiac effects of white snakeroot intoxication in horses. *Equine Practice* 5, 26–32.

Smith, S.M., and Vale, WW. (2006) The role of the hypothalamic-pituitary-adrenal axis in neuroendocrine responses to stress. *Dialogues in Clinical Neurosciences* 8, 383–95.

Snelling, W.M., Cushman, R.A., Keele, J.W., Maltecca, C., Thomas, M.G., Fortes, M.R., and Reverter, A. (2013) Breeding and Genetics Symposium: Networks and pathways to guide genomic selection. *Journal of Animal Science* 91, 537–52

Sparg, S.G., Light, M.E., and van Staden, J. (2004) Biological activities and distribution of plant saponins. *Journal of Ethnopharmacology* 94, 219–243.

Spurlock, D.M., Stock, M.L., and Coetzee, J.F. (2014) The impact of 3 strategies for incorporating polled genetics into a dairy cattle breeding program on the overall herd genetic merit. *Journal of Dairy Science* 97, 5265–74.

Stafford, K. (2014) Sheep veterinarians and the welfare of sheep: No simple matter. *Small Ruminant Research* 118, 106–109.

Stegelmeier, B.L. (2011) Pyrrolizidine alkaloid–containing toxic plants (*Senecio, Crotalaria, Cynoglossum, Amsinckia, Heliotropium,* and *Echium* spp.). *Veterinary Clinics of North America: Food Animal Practice* 27, 419–428.

Stegelmeier, B., and Panter, K. (2012) Poisonous plants and plant toxins that are likely to contaminate hay and other prepared feeds in the western United States. *Rangelands* 34, 2–11.

Stegelmeier, B., James, L., Panter, K., Ralphs, M., Gardner, D., Molyneux, R., and Pfister, J. (1999) The pathogenesis and toxicokinetics of locoweed (*Astragalus* and *Oxytropis* spp.) poisoning in livestock. *Journal of Natural Toxins* 8, 35–45.

Stegelmeier, B.L., Davis, T.Z., Green, B.T., Lee, S.T., and Hall, J.O. (2010) Experimental rayless goldenrod (*Isocoma pluriflora*) toxicosis in goats. *Journal of Veterinary Diagnostic Investigation* 22, 570–577.

Steinfeld, H., Gerber, P., Wassenaar, T., Castel, V., Rosales, M., and de Haan, C. (2006) *Livestock's long shadow: Environmental issues and options.* FAO, Rome, Italy.

Stickel, F., and Seitz, H.K. (2000) The efficacy and safety of comfrey. *Public Health Nutrition* 3, 501–508.

Strickland, J.R., Looper, M.L., Matthews, J.C., Rosenkrans, Jr., C.F., Flythe, M.D., and Brown, K.R. (2011) St. Anthony's fire in livestock: Causes, mechanisms, and potential solutions. *Journal of Animal Science* 89, 1603–1626.

Stubsjøen, S.M., Flø, A.S., Moe, R.O., Janczak, A.M., Skjerve, E., Valle, P.S., and Zanella, A.J. (2009) Exploring noninvasive methods to assess pain in sheep. *Physiology and Behavior* 98, 640–648.

Tokarnia, C.H., Döbereiner, J., and Peixoto, P.V. (2002) Poisonous plants affecting livestock in Brazil. *Toxicon* 40, 1635–1660.

Tortelli, P.F., Barbosa, J.D., Oliveira, C.M.C., Dutra, M.D., Cerqueira, V.D., Oliveira, C.A., Riet-Correa, F., and Riet-Correa, G. (2008) Intoxicação por *Ipomoea asarifolia* em bovinos e ovinos na Ilha de Marajo. *Pesquisa Veterinária Brasileira* 28, 622–626.

Umstatter, C. (2011) The evolution of virtual fences: A review. *Computers and Electronics in Agriculture* 75, 10–22.

Uwai, K., Ohashi, K., Takaya, Y., Ohta, T., Tadano, T., Kisara, K., Shibusawa, K., Sakakibara, R., and Oshima, Y. (2000) Exploring the structural basis of neurotoxicity in C_{17}-polyacetylenes isolated from water hemlock. *Journal of Medicinal Chemistry* 43, 4508–4515.

Van Kampen, K.R., and James, L.F. (1972) Sequential development of the lesions in locoweed poisoning. *Clinical Toxicology* 5: 575–580.

Van Reenen, C.G., O'Connell, N.E., Van der Werf, J.T., Korte, S.M., Hopster, H., Jones, R.B., and Blokhuis, H.J. (2005) Responses of calves to acute stress: Individual consistency and relations between behavioral and physiological measures. *Physiology and Behavior* 85, 557–570.

Van Reenen, C.G., Van der Werf, J.T., O'Connell, N.E., Heutinck, L.F., Spoolder, H.A., Jones, R.B., Koolhaas, J.M., and Blokhuis, H.J. (2013) Behavioural and physiological responses of heifer calves to acute stressors: Long-term consistency and relationship with adult reactivity to milking. *Applied Animal Behaviour Science* 147, 55–68.

Veissier, I., and Miele, M. (2014) Animal welfare: Towards transdisciplinarity – the European experience. *Animal Production Science* 54, 1119–1129.

Veissier, I., Butterworth, A., Bock, B., and Roe, E. (2008) European approaches to ensure good animal welfare. *Applied Animal Behaviour Science* 113, 279–297.

Verdes, J.M., Moraña, A., Gutiérrez, F., Battes, D., Fidalgo, L.E., and Guerrero, F. (2006) Cerebellar degeneration in cattle grazing *Solanum bonariense* ("Naranjillo") in western Uruguay. *Journal of Veterinary Diagnostic Investigation* 18, 299–303.

Villalba, J., and Provenza, F. (2007) Self-medication and homeostatic behaviour in herbivores: Learning about the benefits of nature's pharmacy. *Animal* 1, 1360–1370.

Villalba, J., Provenza, F., and Manteca, X. (2010) Links between ruminants' food preference and their welfare. *Animal* 4, 1240–1247.

Vollenweider, F.X., Vollenweider-Scherpenhuyzen, M.F., Bäbler, A., Vogel, H., and Hell, D. (1998) Psilocybin induces schizophrenia-like psychosis in humans via a serotonin-2 agonist action. *Neuroreport* 9, 3897–3902.

Wallace, R.J. (2004) Antimicrobial properties of plant secondary metabolites. *Proceedings of the Nutrition Society* 63, 621–629.

Waldron, D.F., Taylor, C.A. Jr., Walker, J.W., Campbell, E.S., Lupton, C.J., Willingham, T.D., and Landau, S.Y. (2009) Heritability of juniper consumption in goats. *Journal of Animal Science* 87, 491–5.

Weary, D.M., Niel, L., Flower, F.C., and Fraser, D. (2006) Identifying and preventing pain in animals. *Applied Animal Behaviour Science* 100, 64–76.

Welch, K., Panter, K., Gardner, D., Green, B., Pfister, J., Cook, D., and Stegelmeier, B. (2008) The effect of 7, 8-methylene-dioxylycoctonine-type diterpenoid alkaloids on the toxicity of methyllycaconitine in mice. *Journal of Animal Science* 86, 2761–2770.

Welch, K., Panter, K., Lee, S., Gardner, D., Stegelmeier, B., and Cook D. (2009) Cyclopamine-induced synophthalmia in sheep: Defining a critical window and toxicokinetic evaluation. *Journal of Applied Toxicology* 29, 414–421.

Welch, K.D., Green, B.T., Panter, K.E., Gardner, D.R., Pfister, J.A., and Cook, D. (2014) If one plant toxin is harmful to livestock, what about two? *Journal of Agricultural and Food Chemistry* 62, 7363–7369.

Wemelsfelder, F., and Mullan, S. (2014) Applying ethological and health indicators to practical animal welfare assessment. *Revue Scientifique et Technique (International Office of Epizootics)* 33, 111–120.

Wiedenfeld, H. (2011) Plants containing pyrrolizidine alkaloids: Toxicity and problems. *Food Additives and Contaminants: Part A* 28, 282–292.

Wina, E., Muetzel, S., and Becker, K. (2005) The impact of saponins or saponin-containing plant materials on ruminant production– a review. *Journal of Agricultural and Food Chemistry* 53, 8093–8105.

Witte, S.T., and Will, L.A. (1993) Investigation of selenium sources associated with chronic selenosis in horses of western Iowa. *Journal of Veterinary Diagnostic Investigation* 5, 128–131.

Ziska, L.H., Emche, S.D., Johnson, E.L., George, K., Reed, D.R., and Sicher, R.C. (2005) Alterations in the production and concentration of selected alkaloids as a function of rising atmospheric carbon dioxide and air temperature: Implications for ethno-pharmacology. *Global Change Biology* 11, 1798–1807.

Figure 6.1: The wolf (*Canis lupus*) is one of the most widely distributed and most controversial of the large predators competing with humans. However, worldwide data indicate annual loss of livestock to wolves averages less than 1% of total livestock numbers. Photo credit: John Laundré.

Figure 6.2: The American cougar (*Puma concolor*) is considered by some to be a significant predator on livestock in the western United States of America. However, data indicate its impact is insignificant in numbers and economic loss. Photo credit: John Laundré

Figure 6.4: Coyotes (*Canis latrans*) are a common medium sized predator in western North America. Though it is the major predator on sheep in this region, annual sheep loss to this predator is only 3% of the total number of sheep raised. Photo credit: John Laundré

Figure 6.5: Eurasian lynx (*Lynx lynx*) are considered to be a major predator on livestock in Europe but losses in most areas account for less than 1% of the total number of livestock. Photo credit: John Linnell.

Figure 6.6: African wild dog (*Lycaon pictus*) is one of a suite of medium to large predators known to prey on livestock in Africa. However, the losses of livestock from this and the other species are less than 1% of total number of livestock. Photo credit: John Laundre

Figure 6.7: American jaguar (*Panthera onca*) is the major large predator in Central and South America. Losses to the livestock industry in these regions are normally less than 2% of the total number of livestock. Photo credit: John Laundré

chapter six

Predation

John Laundré

INTRODUCTION

Webster defines predation as: "the act of killing and eating other animals, the act of preying on other animals." Based on this definition, the focus of this chapter would seem quite clear: the killing and eating of domestic livestock raised in extensive agricultural systems. However, predation on livestock is far from a clear and simple subject. It is an issue of many facets and often injected with more emotion than science. It is one thing for a predator to kill a wild deer, one we *might* have hunted for ourselves. It is indeed another thing when that predator kills an animal that we were specifically raising for our use.

In analyzing the issue of predation on livestock, I will first provide a brief history of human pastoral coexistence and conflicts with predators, which in itself is a rich and fascinating one and could fill volumes. I will then fast forward to modern time and analyze the extent of the problem of predation today.

This analysis will primarily be focused on extensive agriculture in the United States, mainly because this is where the most complete data exist on the impacts of predators on livestock operations. However, I will provide comparative data on predation in extensive systems from other regions of the world to see if the data from the United States is an anomaly or typical of the rest of the world. The questions to answer are just how much of a problem is predation on livestock and what are the ways in which this predation can be reduced?

A BRIEF HISTORY: THE BOY WHO CRIED WOLF!

Our antagonistic relationship with other predatory species existed long before we domesticated animals. In fact, how we interacted and viewed these other species probably helped to define that relationship once predators started killing and eating those domesticated animals.

State University of New York at Oswego, USA

Before domestication, we were a hunter-gatherer species much like every other animals species on the planet. For the hunting part, we relied on a wide variety of animal species from small to big. Not surprisingly, many of these species we preyed upon were also the prey of other predators, many of them large and fierce in nature. In fact, some probably also viewed us as prey! Because of these two aspects – they ate what we did *and* they sometimes ate us – we very early on developed the two basic attitudes we still hold today: Other predators are competitors and they are threats to us. If they were not killing and eating animals *we* could have killed and eaten, they were predisposed to killing and eating *us*! Consequently, from these early experiences, just as with wolves (*Canis lupus*) and cougars (*Puma concolor*) of today, competition and fear have driven our relationships and conflicts with other predators. See Fig 6.1 in plate section.

Once we did domesticate animals, around 8,000 BC (Zeder, 2008), the conflict only intensified. As we found raising animals under controlled conditions (corrals [intensive] or herding [extensive]) made it more efficient and easier for us to obtain animal protein, so it did for the other predators! Add to this the fact that domestication decreases the "wildness" of animals (for ease of handling). Imagine a predator, as we had to before domestication, having to scour the landscape for animals that were both dispersed and wary and then coming across a lot of unwary ones all in a bunch! It indeed became a situation that was all too easy to not resist. Thus, the antagonistic relationship between humans and other predators arose. It is of interest that in native societies, e.g., North America,

where animal husbandry was not well developed, intense dislike of other predators was not as strongly developed, even though humans were occasionally killed by them. This further indicates that the antagonistic relationship we developed with other predators is more closely related to the competition aspect, specifically regarding domestication of animals.

Given this threat from predators on livestock, it is not surprising that early on, humans began developing methods of trying to control or reduce the loss of animals to predation. Probably one of the earliest inventions following domestication was the development of fences, corrals, to not only keep domestic animals in but wild predators out! This obviously led to the varied intensive agricultural practices, which are not the subject of this chapter. Suffice to say, one can find predation risk reasons behind the many intensive systems that exist today. There is a reason European farmers build "housebarns" directly connected to their stables (Tishler and Witmer, 1986).

Our task on the other hand, is to look at the extensive management systems. By definition, this refers to the practice of placing livestock outside of protective buildings or fences. Why would we do this, knowing the danger of predators? Interestingly, most of the extensive management developed in what could be termed marginal lands (Huntsinger and Starrs, 2006). To intensively care for livestock, it is necessary to have sufficient fodder nearby to gather and bring to the animals. In marginally productive lands, it becomes easier to bring the animals to the widely dispersed fodder and let them look for it. We see this rationale still today where livestock are allowed to roam in the low productive mountainous lands of western US but are more restricted in the East, where abundant forage can be gathered.

Though the above rationale for extensive livestock management might be a little too simplified, suffice to say allowing livestock to roam over the landscape in search of food was and still is a common practice. It is a practice, however, that increases the conflict between humans and predators. Not surprisingly, then, early on in the use of extensive management, measures were taken to reduce the conflict. The most commonly used method involved the use of herders to travel along with the flock or herd. The role of the herder was twofold. First, they were responsible for moving the animals around the landscape so as to find sufficient food for them. The second was to provide protection, or as in the case of the boy who cried wolf, as an alarm system to alert others of imminent danger from predators. Being a herder, then, has been and is even today, an important element of extensive management systems. From the shepherds of Bethlehem to the cowboys and shepherds of the old, and not so old, West, putting humans between the flock and the predator was the main method used to reduce the threat of predation on livestock. In a sense it was a system of extensive management but with intensive monitoring.

A second method of dealing with predation risk was simply to remove the predators. This approach was not always effective in the times before more modern control methods. Traps and spears and arrows might have been effective for immediate predation threats but because of the low human density and the large wild areas available (typical

of areas of low productivity), predator removal was on a localized scale. It was not until the development of modern firearms and poisons that humans were able to exert control on predator populations over broad landscapes. This brings us to the next section, predation and livestock today, where, using the US as an example, we look at how we preemptively removed predators on a continental scale.

PREDATION AND LIVESTOCK TODAY

As mentioned before, the indigenous peoples of North America coexisted with predators with little conflict. Of the larger predators, bears, wolves, and cougars, only people living in the western part of the continent had to worry about safety from grizzly bears (*Ursus arctos*) as wolves and cougars were never and still are no meaningful threat. As there were no domestic animals, there was also no concern in that regard. There appeared to be abundant native ungulates and other species used for food so that competition for resources between humans and predators also appeared to be low. What existed, based on the many stories and legends passed down, was a tolerance for, even a respect of, predators as part of the system the indigenous people lived in and shared with predators (Pierotti, 2011).

Of course that all changed when Europeans arrived on the shores of the Americas. The humans that arrived brought with them domestic livestock and centuries-old fears and biases regarding predators in general and large predators in particular. As mentioned above, fear and competition drove the European view of predators and those views quickly established themselves in the early settlers and advanced across North America as descendants of the first immigrants colonized westward. As vast open areas were available, extensive management of livestock was a common practice, especially as these European descendants moved into the open western plains areas.

Besides introducing domestic livestock to the land and letting them roam freely, the new inhabitants decimated native populations of prey for food and clothing (Anonymous, 2014) and through habitat destruction (Tchir et al., 2014). This massive hunting of wildlife, often on a commercial basis, swept across the continent and was even encouraged by the government, for example removal of bison (Smits, 1994). Not surprisingly, as native prey began to dwindle, native predators turned more to domestic prey, exacerbating the conflict with humans.

In efforts to reduce predation, real or imagined, on livestock – and themselves – the new occupants chose preemptive removal of predators as the only option. Early on in colonial history, people were encouraged, even paid, to kill predators (Penna, 1999), regardless of whether they were causing problems or not. The official stance was that all predators, large and small, were undesirable and systematic removal of them began, with some of the first bounties being paid for their removal instated as early as 1630 (Penna, 1999). As with the view expressed regarding indigenous people, the only good predator was a dead predator.

And the war began, one that continues today. Initially, the efforts were limited in scale and in techniques. Colonialists only had what are now considered primitive firearms of limited range and accuracy. An interesting side note is that before the development of the rifled barrel, the early guns were of such limited accuracy that, once leaving the barrel, the rounded ball could veer off in any direction. As the accuracy of firearms, and also trapping equipment, improved, the effectiveness in killing predators, and all wildlife, followed suit.

This strategy of complete removal worked effectively in the Eastern US with the removal of the last wolves and cougars occurring in the late 1800s to early 1900s (Whitaker and Hamilton, 1998; Penna 1999). However, even before that time, these animals had been effectively removed from the more settled lands, only clinging to survival in more remote areas, where they were also eventually sought out and killed. This pattern of society-sponsored removal of predators moved westward as European descendants pushed across the country. With the discovery of strychnine in the early 1800s, a new weapon was quickly adapted to the fight (Fagerstone et al., 2004) – chemical poisons. Baiting carcasses with strychnine effectively removed grizzly bears from all but a few remote mountainous areas and wolves from 48 of the 49 now recognized states. Cougars fared better because of their reluctance to eat carrion and their secretive nature. However, they were still eliminated from the eastern two thirds of the United States, persisting, again, only in the more remote mountainous areas of the West. It is of interest that the removal of the last grizzly, wolf, or cougar was often heralded by many at the time, even as now, as a success, making the landscape safe for us and our livestock. Thus we often have exact dates of when the "last" wolf, cougar, or grizzly was killed.

Though not viewed as a threat to us or even to larger livestock, the expansion of the sheep industry in the West exasperated the conflict with coyotes (*Canis latrans*). Equal, if not more intensified, efforts to remove coyotes from the landscape, however, have not proven as successful. After decades of shooting and poisoning, coyotes are not only still doing well across the western states but have expanded eastward to the coast, filling the vacancy left by the removal of wolves and cougars (Mastro, 2011).

Although the removal of wolves and grizzly bears and the forcing of cougars to the distant corners of the landscape was heralded by many as a victory, there were many others who saw this as a tragedy. Though the killing of these animals was accomplished during our manifest destiny period, new evidence from the growing science of ecology indicated that instead of a victory, removal of these predators was an ecological loss. One of the earliest to sound the alarm was Aldo Leopold, considered by many as the father of modern wildlife conservation. Based on his experiences in the West, especially with regards to the Kaibab Plateau in Arizona, Leopold began to question the commonly held belief that the only good predator was a dead predator. As with the Kaibab, elimination of predators resulted in an explosion in the number of native ungulate species, substantial habitat destruction, and eventually massive die-offs (Leopold, 1943; Leopold et al., 1947; Binkley et al., 2006). These results were fortified by additional examples such as moose on Isle Royale (Nelson et al., 2010) and, today, with deer

in the forests of the east (Horsley et al., 2003). Leopold argued that predators were needed on the landscape as a stabilizing force, keeping nature in balance. As he stated: "I now suspect that just as a deer herd lives in mortal fear of its wolves, so does a mountain live in mortal fear of its deer." His voice joined earlier ones such as Henry Thoreau, John Muir, and others, and began to awaken American society, too long shaped by fear and ignorance of predators, to a more enlightened view of predators and their role and importance in nature.

Today, those original ideas expressed by these pioneers in ecology have elevated the status of predators from vermin to essential keystone species in ecosystems. As keystone species, those whose positive role far outweighs their numbers, large predators such as wolves, grizzly bears, and cougars, are viewed by many as shepherds of the large ungulates (Laundré, 2012). As with human shepherds, they keep the herds they tend culled to healthy numbers and moving around the landscape to prevent overuse and destruction.

With this enlightened view of predators, laws were enacted, for example the Endangered Species Act for wolves and grizzly bears and state game status for cougars, that were designed to reduce the uncontrolled killing and, in the case of wolves, actively restore populations. These actions have been successful and today we see all three of these species expanding their ranges significantly. It is interesting that we are seeing a similar increase in wolves, Eurasian bears (*Ursus arctos arctos*), and Eurasian lynx (*Lynx lynx*) (Stahl et al., 2001; Rigg et al., 2011; Wagner et al 2012) in Europe as people's attitudes change toward these species. However, though these could all be considered ecological successes, they have raised age-old fears from the livestock industry. After several generations of not having to think about large predators, faced with their return, ranchers and herders are reluctant to accept this new threat, perceived or real, to their extensive management of livestock. And too, there is always the persistent "problem" with coyotes in the United States.

So the war rages on, even more intensified with the reintroduction of wolves in the West. Each year the United States continue to spend more than US$100 million per year in "Wildlife Services" to kill hundreds of thousands of predators to protect our livestock. But is it a necessary war? For the safety of our livestock and ourselves, do we need to remove all predators from the landscape yet again? And if we do, how do we reconcile this with the ecological role we now know predators play? For example, current "management" strategies in the western states, where wolves have returned, are to reduce wolf populations to "relic" status, levels well below their ecological effectiveness. If we have the scientific evidence of their importance in ecosystems, such draconian measures, short of elimination, still basically removes them from the landscape. To reconcile livestock and ecological concerns, unlike the old days, the question becomes not how we can get rid of or reduce predators to relic status, but how we can reduce their impacts on livestock while maintaining ecologically viable predator populations – coexistence. This I will address further on in this chapter. But first we need to determine just how much of a problem having predators on the landscape is to

the extensive management of livestock. This helps us to evaluate if coexistence is possible and if it is, then we can assess how to do it.

In assessing the impact of predation on extensively raised livestock, it becomes basically an analysis of numbers; numbers of animals killed and economic value of those animals. We need to objectively look at just how much economic damage predators do to the livestock industry. We need to look at not only total numbers, animals and dollars, lost but we need to look at percentages, the percentage of these losses to the total number and economic value of all livestock exposed to predation. Additionally, we especially need to compare losses from predation to other losses of animals and dollars. These become critical comparisons relative to predation impacts on extensive management of livestock because they allow us to evaluate the relative importance of predation on this practice and just how we should handle it. These comparisons are the topic of the next two sections where I look first at the lethal impacts of predators on livestock – just how many are they killing? I will then look at the non-lethal impacts, that is, does having predators on the landscape have an economic effect?

LETHAL IMPACTS OF PREDATION ON LIVESTOCK

First of all, let's state the obvious: Predators kill animals, it is how they make their living; it is part of the ecological role they have had for millennia. Consequently, predation in itself is not bad, predators are not bad, it is just an ecological fact. However, it is difficult to remove the emotion of a rancher seeing an animal that he or she has raised being killed by a predator. But remove the emotion we must. One of the tenets of the North American Model for Wildlife Conservation is that science, rather than emotion, should guide us in managing wildlife, including native predators. Just as we should not let the "emotion" of people who may not like hunting sway decisions and policy, we should not let the "emotion" of people who might experience losses of domestic animals to predators sway decisions and policies on predator management. We must let the facts speak for themselves. If those facts indicate that, like the emotional feelings, predation is a serious economic threat to extensive livestock practices, then that leads us to a certain set of management decisions. If, however, the facts indicate that predation on livestock is not a serious economic threat, then we need to incorporate that into our management decisions.

So the first thing to look at are the numbers! Just how serious is predation to the livestock industry? What then, are the facts, the numbers behind livestock predation? First, I will separate out large livestock, cattle, from the medium-sized ones, sheep and goats. I will not look at smaller animals such as chickens and ducks as they are rarely raised in an extensive manner. I will not consider horses here either as most would admit that horses, though cherished by many and used economically, are not essential to the economics of the livestock industry. I will primarily concentrate on the 11 western states of the United States (AZ, CA, CO, ID, MT, NM, NV, OR, UT, WA, WY) as that is where cattle, sheep, and

sometimes goats, are most commonly raised extensively. It is also where the primary predators, coyotes, cougars, and now wolves, are found. See Figure 6.2 in the plate section. I have not included Texas or Oklahoma as neither have wolves and only Texas has a limited regional population of cougars. To include these and more eastern states where cougars and wolves are not common would tend to dilute the impacts of these predators.

First, I will examine how many head of livestock are raised in the Western US. I will then look at where the loss of animals occurs from all sources and then compare the loss of livestock from predators to the total number of animals raised (total predation rate) and then to the other livestock losses (cause-specific predation rate; Joly et al., 2009). All the data used are from the annual inventory and records published by the USDA (http://www. nass.usda.gov/Statistics_by_Subject/ index.php?sector=ANIMALS%20&%20 PRODUCTS). The latest summary on predation losses is from 2010 and so production inventory levels, number of livestock, for January of that year are also used.

Cattle

In considering cattle, I will limit the analysis further to beef cattle, again, as these are the animals primarily raised in extensive management systems. To have a starting number, the total number of cattle, calves included, in the US in January 2010 was 93.7 million head. Of these, 18.9 million head are raised in the 11 states in question, representing approximately 20% of total US production (Figure 6.3). Of these 18.9 million head, 2.1 million (11.1%) were listed as being in feedlots, reducing the total to 16.8 million head. In each state, a different percentage of cattle are dairy cattle, presumed to be raised in a more intensive manner. Calculating the percentage of dairy cattle for each state and subtracting those numbers from the total number of head, leaves us with approximately 9.8 million head (51.8% of the total number of cattle in these states) raised in an extensive manner (range animals) and thus, susceptible to predation. Approximately 46.6% of these animals would be calves (4.6 million head). Relative to total US cattle production then, approximately 10.4% of the total number of animals would be exposed to predation risk in these 11 states.

Now that we have the number of head of cattle "out there," let's look at what happens to them. Again, the numbers come from the USDA publications. I will assume that predation numbers for the western states apply only to the 9.8 million range animals.

As point of reference, across the total 48 states, in 2010, approximately 4.0 million head were lost with 219,900 being killed by all kinds of predators (Figure 6.3). This represents a lost economic value of approximately US$98.5 million. In contrast, all other non-predator losses totaled 3.8 million head and a lost economic value of US$2.3 billion. Thus overall, cause-specific predation losses represent 5.5% of the losses of cattle and about 4% of the financial loss. The overall predation rate (based on 93.7 million head) is 0.2%. Other losses represent 4.0% of total production.

Of course, we are figuring in here the millions of animals that are raised intensively or are in states where they are not susceptible to predation by wolves or cougars. Do these numbers change when we look at these more predator-prone states? Total losses of cattle in the 11 states under consideration were 804,000 head in 2010. The overall figures show that, like the national figures, cause-specific mortality from predation was 6.8% (55,000 head) with non-predator losses again being the majority of cause-specific mortality (93.2%; Figure 6.3). The overall mortality rate from predators was 0.5%, again similar to the national average. The overall loss rate from other mortality causes was 7.6%, several percentage points higher than the national average.

Regarding economic losses, total economic loss from death of cattle for the 11 states was US$488.2 million. Predation losses, valued at US$27.0 million represented 5.5% of those losses. Again, non-predatory losses (US$461.2.5 million) represented 94.5%. Regarding the total value of all range cattle in these 11 states, predator losses represented 0.3% while non-predatory losses represented 6.3%. As with the rest of the nation, economic loss from predation was low compared to losses from non-predatory causes and insignificant (less than 1%) regarding the total value of the western extensively managed cattle industry.

What these numbers indicate is that, contrary to popular emotion-based opinion, the western states are not experiencing exceptionally higher predation rates or economic losses than other states. In fact overall, like the rest of the country, predation rates, cause-specific or total, are extremely low and economic losses are a fraction of the total economic losses suffered by cattle operations.

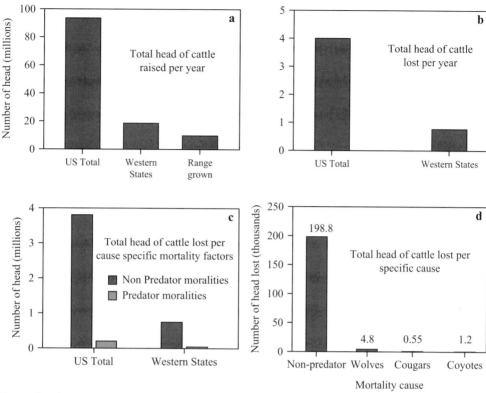

Data are from the USDA National Agricultural Statistics Service http://www.nass.usda.gov/

Figure 6.3: a, Total head of cattle raised (intensively and extensively) in 2010 in the US and the 11 western states under consideration plus the estimate of the number of cattle raised extensively in these states; **b,** Estimate of total head of cattle lost in 2010 in the US and the 11 western states; **c,** Estimate of the total number of cattle lost in 2010 to non-predatory and predatory factors for the US and 11 western states; **d,** Number of cattle lost in 2010 in the three states that had wolves (Montana, Idaho, and Wyoming) to non-predatory causes and to the three main predators, wolves, cougars, and coyotes

Even though predation on extensively managed cattle is low, it is worth the time to look at just how much the three main predators of concern, wolves, cougars, and coyotes, contribute to those losses. A tremendous effort is being make to reduce wolves in the states where they have returned (Montana, Idaho, and Wyoming) and so we need to look at the numbers to see if this effort is worth it. Also, across all the 11 states, pressure is rising to reduce cougar and coyote numbers for the same reason. What

are the predation rates, total and cause-specific, for these three predators?

For wolves, in Idaho, Montana, and Wyoming, in 2010, when they were still protected and close to their highest numbers, wolf predation represented 30.9% of the cause-specific predation mortality. This is 30.9% of the 7.1% predation-specific mortality for these three states. Consequently, the cause-specific mortality from wolves in these three states was 2.2%. This compares to 92.9% non-predation cause-specific for these three states. As for numbers, total cattle mortality from wolves in 2010 was 4,843 head (Figure 6.3) out of 4.5 million head or a 0.1% total predation rate. Non-predatory losses for the three states was 198,800 head or 4.4% total loss rate, almost 50 times higher than losses from wolves. Similar calculations for cougars and coyotes over the ten western states come up with comparable results. To avoid being drowned in numbers, the basics are: Total mortality from coyotes was 1,259 and from cougars it was 555 head (Figure 6.3). Cause-specific mortality from coyotes is 2.0%, and for cougars it is 1.2%. Total predation rate for coyotes was 0.02% and for cougars, 0.01%.

These are the numbers for cattle. Do they differ for sheep? Let's turn our attention to the extensive management of sheep and look at the numbers.

Sheep

As before, I will restrict the analysis to the 11 western states I used for cattle. I will also limit this discussion to just sheep, excluding goats. The reason for this is that sheep are the principle medium size livestock raised in these states and to include goats would further immerse us in numbers. I will also not go into as much detail as with cattle and just "cut to the chase" by looking at the overall numbers. The year used in this case is 2009 as that is the latest predation data available.

To begin with, some basic numbers: 5.6 million sheep in the US, 2.8 million of them (50%) in the 11 western states. We will assume most of these animals are managed in extensive systems and thus vulnerable to predation. Cause-specific mortality from predation nationwide was

38.9% for predators (247,000 head) and 61% (387,000 head) for non-predator causes. Total loss rate from predators was 4.4% and from non-predator causes, 6.9%. For the western states, predation cause-specific mortality was 53.4% (119,700 head) and from non-predatory causes, 46.6% (104,300 head). Overall mortality rate from predation was 4.3% and from non-predatory causes, 3.8%. These numbers indicate that predation rates in the western states were similar to the rest of the country.

Unfortunately, we do not have predator-specific mortality data available for all the western states for sheep. However, data are available for five states (ID, MT, CO, UT, and WY). As three of these states are where wolves are now found, along with cougars and coyotes, I will use these states to look at the impact these three main predators have on the sheep industry. I will also include dogs, as data indicate that they are a major predator on sheep. First, cause-specific mortality from all predators for these five states averaged 49.1%, slightly less than the western average. Cause-specific non-predator mortality was 50.9%, approximately the same as for predators. Overall predation rate for these states was 4.3% and 3.8% for non-predator causes. These numbers are similar to the rest of the West, indicating that we can use these states to look at predator species-specific trends.

Coyotes are the largest cause-specific mortality among predators, averaging 65.7% of all predator-related mortality. See Figure 6.4 Appendix 1 page XXX). Dogs account for 3.0% while cougars are 5.0% and wolves (for ID, MT, and WY) are at 7.0%. Although coyotes accounted for the vast majority of predation-related sheep mortality, the overall rate of sheep mortality from coyotes is 3.0% with cougars and wolf overall predation rates being 0.2%.

What do all these numbers tell us? To use the metaphor of the boy and the dike, we are trying to prevent a loss from a trickle while ignoring the massive breach in the dike. The economic loss for cattle from all predators in the eleven western states was US$24.5 million while losses from other sources were US$441.5 million, 18 times higher. Although for sheep, the cost of predation is closer to losses from other sources, economically, non-predator losses (US$36.3 million) exceeded predator losses (US$20.5 million). However, because predation is an emotional issue, this unfortunately leads to decisions based on emotion and not science or even sound economics! For example, the state of Idaho has allocated US$2 million to "control" wolf predation. In contrast, the economic loss (cattle and sheep) from wolves was approximately US$1.5 million and the economic losses from other causes was US$56.4 million dollars. Surely, there would be a better return on the investment if that US$2 million was spent trying to stem the loss of the US$56.4 million.

What all these numbers tell us is that we are wasting our time and money on losses of livestock from predation, specifically from wolves, cougars, and coyotes, while there are more serious threats to the economic viability of extensive cattle and sheep ranching. If states want to spend money, they should consider supporting programs that would reduce the real threats to livestock; predation is not one of these.

Many of these non-predatory causes are easily preventable by improving animal husbandry practices. In fact, much of the predation-caused mortality can also be prevented by improved husbandry practices, further reducing these already insignificant losses. How to reduce those losses is the topic discussed later. First, I would like to look at global trends in depredation rates.

GLOBAL TRENDS IN DEPREDATION RATES

I have given a fairly detailed analysis of depredation losses from predators on livestock in the United States of America. The result of this analysis has shown that depredation rates compared to total livestock raised in extensive management systems is relatively low (0.2% for cattle, 4.3% for sheep). How do these numbers compare to other parts of the world? Is the United States an anomaly or are depredation losses in other continents similar?

Let's first consider Europe where, for the wolf at least, we are considering similar predators. Additionally, the Eurasian lynx can be considered an analog to the American puma. (See Fig 6.5 in plate section). For both species, as with wolves in North America, there are established populations as well as areas where they have recently recolonized (Stahl et al., 2001; Rigg et al., 2011). When we look at reported studies of depredation losses for various countries, annual losses of sheep and cattle in Italy from wolves is 0.4% (cattle) to 0.7% (sheep) of total herd numbers (Gazzola et al., 2008). In France, regional sheep losses from lynx were from 0.14–0.59% of the total number of sheep (Stahl et al., 2001). In Slovakia, sheep losses in 2004 from Eurasian bears and wolves were 0.2 and 0.8%, respectively, of total head numbers (Rigg et al., 2011). Zimmermann et al. (2003) reported predation losses by bears in south-eastern Norway as 0.5% for cattle and 5.4% for sheep. Within the eight years after wolves recolonized eastern Germany, predation on livestock in that region was reported to be around 0.6% of total herd biomass (Wagner et al., 2012). The highest predation rates by wolves appears to be Greece (2.3% cattle, 5.8% sheep; Iliopoulos et al., 2009) and Poland (7.1% sheep; Nowak et al., 2005). In most of these studies it was also reported that predation losses were not evenly distributed, with some livestock owners experiencing higher losses than others (Stahl et al., 2001; Nowak et al., 2005; Gazzola et al., 2008; Rigg et al., 2011).

Moving further east, Banerjee et al. (2013) reported Indian livestock owners' predation losses of cattle to lions (*Panthera leo*) of approximately 5%. In the Trans-Himalayan region livestock owners reported annual losses of 2.9% from Tibetan wolves (*C. l. chanco*), snow leopards (*Uncia uncia*), and Eurasian lynx (*L. l. isabellina*) (Namgail et al., 2007). Predation specifically by snow leopards in Nepal was estimated to be 3.9% (Wegge et al., 2012). Wolf predation on livestock in central Iran for 2007–8 was 0.7 to 0.9% of total herd numbers (Hosseini-Zavarei et al., 2013). However, they also reported that almost 6 times more animals died from non-depredation causes. Li et al. (2013) reported for northwest Yunnan in China an annual predation loss of 2.1% of cattle, horses, and donkeys by wolves, Asiatic black bears (*Selenarctos thibetanus*), and dholes (*Cuon alpinus*).

In Africa, Patterson et al. (2004) reported livestock losses to lions near the Tsavo National Park as 2.4% annually. Wild dogs (Fig 6.6 Lycaon pictus in the plate section) in northern Kenya had a depredation rate of 0.007% of total herd numbers of cattle and sheep combined (Woodroffe et al., 2005). In a multiple predator complex of lions, spotted hyenas (*Crocuta crocuta*) and leopards (*Panthera pardus*) adjacent to the Maasai Mara National Reserve, combined predation rates for sheep and cattle were 0.6 and 0.2%, respectively, of total herd numbers (Kolowski and Holekamp, 2006).

Australia is often noted for its problems with livestock and dingos (*Canis lupus dingo*) with its prevalent use of lethal control measures and its infamous dingo fence (Downward and Bromell, 1990; Allen and Sparkes, 2001). Impacts of dingos on the sheep industry have been described as a widespread problem costing millions of dollars a year (Allen and Sparkes, 2001; Hewitt, 2009). It is commonly accepted that dingos are not major predators on cattle, with most of the damage being to the sheep industry. However, there have been few studies quantifying the actual rate of loss due to dingo predation. Of the few published data that exist, the predation rate on adult cattle does appear to be low. Predation on calves has been reported to average 2.8% for Queensland cattle owners (Hewitt, 2009). Predation rates on calves from 9% to 32% were reported by Allen (2014). However, the higher predation rates were from an area where dingos were controlled by poison baiting and strongly related to periods of low rainfall (Allen, 2014). In general, Allen (2014) concluded that controlling wild dogs was unnecessary because they normally did not prey on calves.

Relative to sheep, estimates of economic annual losses as high as A$16.9

million have been reported (Hewitt, 2009). However, relative to the percentage of sheep killed by dingos to the total number of sheep raised, mean losses summarized in Fleming et al. (2001) ranged from 0.7% to 1.3%. The discrepancy between the percentage lost and the total economic value is related to the high number of sheep that were being raised. In the original study by Fleming and Korn (1989), reported losses of sheep to dingos from two areas in Queensland were in excess of 4,000 animals, but the total number of sheep in these areas was 15.1 million, making the loss less than 0.01%. Naturally, though the monetary amount lost to dingos is high, the percentage of this economic loss by predation to the total value of sheep raised was also well under 1%.

The last area to consider is Latin America, including Mexico, Central America, and South America. The two main livestock predators in these areas are cougars and jaguars (*Panthera onca*). (See Fig 6.7 in plate section). In northern Mexico cattle losses to cougars in the Chihuahuan Desert were estimated to be 1.9% of total cattle numbers with losses from other causes such as drought, robbery, and diseases being 7 times higher (Bueno-Cabrera et al., 2005). In Costa Rica, predation by jaguar and cougars on mixed livestock, including cattle and horses, was estimated to be 0.5% (Amit et al., 2013). In Brazil, several studies reported cattle losses to pumas and jaguars to be less than 1.0% (Dalponte, 2002; Mazzolli et al., 2002; Conforti et al., 2003; Michalski et al., 2006; Palmeira et al., 2008). Mazzolli et al. (2002) also reported that cattle mortality from other causes assumed a greater importance in herd productivity. Hoogesteijn et al.

(1993) reported calf losses to jaguars and cougars in Venezuela to be 1.3–3%. They also reported that predation mortality on calves represented only 6% of all calf losses or deaths. The one study looking at cougar predation losses to sheep in Brazil reported losses across 16 different ranches to average 24.9% (Mazzolli et al., 2002). As with other regions, losses of livestock from predation were not evenly distributed and often related to habitat, weather, and husbandry factors (Mazzolli et al., 2002; Michalski et al., 2006; Palmeira et al., 2008).

Based on the studies cited from other regions of the world, it appears that the results from the US cattle industry are representative of a global trend. Given the vast difference in predators, habitats, cultures, and husbandry techniques, the similarity of the pattern of relatively low predation losses is interesting. It seems that regardless of the system, overall depredation on livestock is not very high. However, as with the US, mortality losses within each area were highly variable among livestock owners, placing burden of loss unequally on certain individuals. The reasons for this inequity, as in the US, seems to be a combination of the level of husbandry and habitat characteristics of the environment. Although not all the studies reported livestock losses from other causes, at least four studies (Hoogesteijn et al., 1993; Mazzolli et al., 2002; Bueno-Cabrera et al., 2005; Hosseini-Zavarei et al., 2013) reported that non-predator losses were significantly higher, another similarity with the data from the US. Consequently, it appears that depredation rates worldwide are not only low but most likely are not the major loss of livestock raised in extensive management systems. However, because of the emotion involved in what is probably related to the often, what we consider, gruesome manner in which predators kill their prey, the pure numbers will probably not convince most livestock raisers to simply ignore depredation losses. Thus, it becomes important to now discuss the possible options of reducing even further the loss of livestock from predation to a possibly more acceptable level.

REDUCING PREDATION RISK

As mentioned earlier in this chapter, one of the major techniques used by herders of livestock is vigilance. Sheep and cow herders are steeped in our past and current livestock culture worldwide. Just by the various names, in many languages, we can understand the importance of the cowboy, gaucho, vaquero, or shepherd, pastor, protector in the raising of livestock. The last name "protector" more than subtly hints at the real role of people who tended flocks and herds – to protect the animals in their charge from all types of dangers, including predators. In fact, animal husbandry in extensive agricultural systems revolved around, depended upon, humans guarding their investment, their livelihood, the animals under their care.

In addition to the vigilant herder, another common pastoral image is of the herd dog, Since humans domesticated livestock and dogs, the combination of shepherd and dog is legendary. Although often some breeds of dogs were used to drive and manage flocks or herds, others have also played a protective role, again against possible predators. Well before the poet Virgil in 26 BC

Rome wrote of guard dogs, they have been used so that "You'll never fear midnight thieves in the stables, *attacks of wolves* [my emphasis]" (Kline, 2002). All across Eurasia wherever humans herded flocks of livestock, their guard dogs followed.

Needless to say, not all breeds of dogs made good livestock guard dogs. The main characteristics needed for a good guard dog is that it should be large, mild mannered toward humans and livestock, and, of course, aggressive toward predators (Urbigkit and Urbigkit, 2010). In general, the three qualities desired in guard dogs are trustworthiness, attentiveness, and protectiveness. Over the centuries, certain breeds have been developed that fit this bill for livestock protection dogs (LPD). Common breeds known today include the Pyrenees, Komodor, and Maremma (Urbigkit and Urbigkit, 2010). However, the list of LPD breeds is long because they were independently developed over a large geographic area, again indicating their importance and effectiveness. Each has their individual traits that make them well suited for the areas for which they were developed. Two general patterns of guarding can be found in LPDs. In the first, the dog follows the flock or herd closely and wards off predators if they come. In the second, the dog patrols the periphery of the herd, keeping predators away from the livestock (Green and Woodruff, 1993). Although many breeds have been used historically as LPDs, individual animals do need to be trained before they can be used to protect herds or flocks. The process of training dogs as LPDs has been documented in various places. Coppinger and Coppinger (2007) provide an overview of the process plus links to the diverse literature available for training of LPDs.

How do they protect the herd? Although guard dogs will show aggression toward predators, they rarely will fight them to the death. Instead, they use standard canine social dominance behaviors commonly used in aggressive interactions, for example growling/barking, threatening postures, and as a last resort, physical contact against approaching predators (Coppinger et al., 1988). It appears that once a guard dog has established its dominance over predators, the incidences of conflict reduce and predatory animals often leave the immediate area.

Are they affective? One would say that since they have been around for several thousand years, a general answer is yes. In Kazakhstan, with one of the highest densities of wolves in the world, guard dogs are considered highly effective in reducing predation and are extremely valued for this skill (Urbigkit and Urbigkit, 2010). Interestingly, it appears that in Eurasia, because guard dogs and predators such as wolves have coexisted for so long, there is evidence that wolves have developed an almost instinctive sense to leave flocks of livestock alone when they are accompanied by guard dogs. More quantitative analyses (summarized in Gehring et al., 2010) indicate that LPDs can be highly effective, up to 100%, but the effectiveness can be variable, depending on the breed and the predator it is facing. Some authors (Green and Woodruff, 1984; Andelt, 1992; Andelt and Hopper, 2000) reported that guard dogs worked well in reducing coyote predation on sheep. Against larger predators, for example wolves and bears, guard dogs can also be effective but they

do run the risk of being killed (Urbigkit and Urbigkit, 2010). Consequently, in many Eurasian countries with wolves, herders often fit their dogs with protective collars to reduce the chance of being injured or killed (Urbigkit and Urbigkit, 2010). However, in their overall analysis, Gehring et al. (2010) concluded that the use of guard dogs can be valuable tools in protecting livestock. This conclusion is consistent with earlier reviews of guard dog effectiveness (Smith et al., 2000; Coppinger and Coppinger, 2007).

If intense shepherding and guard dogs are effective methods of reducing predation on livestock, why is there such a concern, albeit over-emphasized, based on the data above, over predation impacts on the livestock industry? Unfortunately, today, at least in the Western US, herds and flocks are too big and vaqueros and shepherds too few to provide adequate protection. The common practice by ranchers in the Western US is to just turn their cattle loose on range land, mostly federally owned, in late winter (south) or early summer (north) and provide minimal supervision (Personal observation) until they are gathered up later in the fall. Sheep owners commonly are only in the field when it comes to moving their flocks to different allotments, meanwhile leaving them in the care of often only one shepherd (Personal observations). Evidence of this low level of maintenance is that even today in the US over 16,000 head of cattle and sheep per year are stolen and over 490,000 head die per year from unknown, non-predator-related causes. Although guard dogs have been shown to be effective, only 28% of the sheep herders use them to protect their herds (Andelt, 2004). In the three western states currently with wolves (ID, MT, and WY), only 25% of the cattle ranchers use LPDs. Only 15% of the cattle ranchers use intensive herding practices. The conclusion to draw here is that although predation losses only account for 0.4% for cattle and 4.4% for sheep, implementation of more intense herding practices and guard dogs could reduce these impacts even further. This is especially noteworthy regarding sheep, where coyotes are the primary cause-specific mortality and the use of guard dogs seems to be the most effective (Andelt, 2004).

On a worldwide basis, the use of guard dogs has also been shown to be effective in significantly reducing losses by predation (Rigg et al., 2011; Van Bommel and Johnson, 2012; Rust et al., 2013; Tumenta et al., 2013). Again, the historical use of guard dogs from thousands of years ago attests to their utility (Coppinger and Coppinger, 2007). However, even with this long history, guard dogs are not used universally, for various cultural and perceived economic reasons although their use is increasing, especially where large predators have been making comebacks.

An added advantage of using guard dogs or increased herding practices is that they are often what we consider non-lethal methods of controlling predation. The idea is not to remove the predator but reduce its predation on livestock. Why is this an advantage? First, as mentioned earlier, knowing now the importance of predators in ecosystems, removal or even extensive reduction of predators should not be the prime objective. We should be striving for methods that allow coexistence of extensive livestock management with predators. These non-lethal methods provide us

with the mechanism of achieving this coexistence. Second, as recent studies have shown with cougars in Washington State, removal of often resident cougars, in this case from hunting, tends to disrupt the social order and can actually lead to increased cougar–livestock conflicts (Peebles et al., 2013). What is left are younger individuals with less hunting experience, leading them to prey upon livestock more than older, established ones. We also have ample data demonstrating that coyotes in areas of high removal have increased reproduction rates, again producing an abundance of young, inexperienced animals prone to select livestock (Gese, 2005). Thus, non-lethal methods of preventing predation on livestock not only enable keeping predators on the landscape but also maintain the social systems of these species that further reduce predation incidences. These are conclusions that Eurasian herders have drawn long ago but that, unfortunately, still elude their American counterparts.

NON-LETHAL IMPACTS OF PREDATORS – THE ECOLOGY OF FEAR

Until now, we have been addressing what we would call the lethal impact of predators on livestock: they kill them. It is obvious that this is of paramount concern to livestock owners as every cow or sheep killed by a predator is one less animal in their herd. Again, though not a significant proportion of all the animals out there nor of those that do die, those animals killed by predators not only represent animals lost but each of them represents a definable monetary loss to the owner. These are the probable reasons why we put so much emphasis on the lethal effects of predators. Additionally, because many predators leave specific signs when they kill an animal, it is relatively easy to document these events.

Although lethal impacts of predators on livestock are the most obvious, there is a new, growing area of predator–prey relations that indicates predators can have their most profound impact on their prey, not by killing them, but by scaring them. Specifically, we are concerned with what the animal's reaction is to being attacked and what it might learn from that experience, if it survives. It is obvious that when an animal is attacked by a predator, it indeed should be scared, fearful. But it is just as obvious that if that animal is killed there is nothing it could learn from the experience! However, what if that animal escapes this near-death encounter, which, according to a multitude of studies, happens as much as 80% of the time (Nellis and Keith, 1968; Mech et al. 2001)? Or how about the other animals that were nearby when the predator attacked? They too should have been scared momentarily but were probably relieved when the predator passed them by! It is the survivors of these near-death experiences and what they might learn from them that is the emphasis of non-lethal effects.

What can, should, these survivors learn? They obviously should learn to be afraid of their predator. However, they should also learn just how deadly or lethal the predator is and how that lethality is affected by habitat. Each predator species has it strengths and weaknesses regarding their efficiency in killing its prey. Wolves, for example, are more effective in open areas where they can run their prey down (Arjo and Peltscher, 2004; Kuzyk et al., 2004); cougars, on the other hand, prefer areas of cover where they can sneak up on their prey (Laundré and Hernández, 2003; Holmes and Laundré, 2006). Thus, the risk of being killed, the predation risk, is dependent on the type of predator and where that predator is encountered on the landscape. Encounters with predators, then, should not only instill a healthy level of fear but allow the survivors to assess the risk of being killed in the future, predation risk, and incorporate that risk into their daily lives – if they want to survive. It is this development of fear in prey, based on their experiences with their predators where they survive, and how they incorporate those experiences, that is the basis of the new ecological concept of the ecology of fear (Brown et al., 1999). Simply stated, the risk of predation should instill fear in prey that should affect not only the internal physiology (stress) of prey but their external behaviors, specifically how and when they use the landscape they live in (Laundré et al., 2001). This in turn will affect the multitude of ecological interactions these prey have with their environment and with other species. Because it only affects the prey that are still alive after an encounter with a predator, we refer to this as the non-lethal impact of predators. How powerful is this non-lethal effect? A predator can only kill an animal once but it can scare it and all the others around it many times, and keep them scared. If the survivors don't learn to be afraid and respond accordingly, they will not survive long.

As indicated, the ecology of fear has an internal component, stress, and an external one, how animals adjust their behavior under the fear of predation. The impacts of this fear can be as subtle as affecting the health and reproductive potential of these animals (Creel et al., 2007; Sheriff et al., 2009, 2010) to altering their use of the habitats of an area based on fear: the landscape of fear (Hernández and Laundré, 2005; Laundré et al., 2010). Increasing numbers of studies of wildlife are demonstrating that fear is a factor to be reckoned with and prey are responding accordingly. What we need to ask now is whether this ecology of fear can have a negative effect on domestic animals raised in extensive grazing conditions.

Do domestic livestock show fear of the predators they encounter? Does that fear affect them physiologically and behaviorally to the point that it can reduce their health or their grazing efficiency, for example how they use the landscape. Both of these factors could have subtle impacts on the survival and productivity of livestock and represent potential losses to the producer beyond just those animals being killed. Basic questions regarding these non-lethal impacts need to be addressed. For example, do lambs/calves grow slower or have lower survival under the stress of predation to themselves or their mothers? Additionally, are some of those non-predator deaths, for example weather, due to predator-induced stress? Or does the fear of predation alter how livestock use an area sufficiently to reduce their foraging efficiency? Answering these and many more questions related to the non-lethal impacts of predation will help more fully understand the true costs to the producer in coexisting with predators on the landscape.

Unfortunately, the concept of non-lethal, fear-induced predator impacts is relatively new and little information is available for domestic livestock under predation risk. This research and knowledge gap greatly limits our current assessment of how non-lethal predation effects might affect the welfare of livestock in extensive conditions. There are, however, some studies of wild prey species that can give us insights into the non-lethal impacts of predation. What follows is a brief review of some of the physiological and behavioral effects of predation risk on wild prey. From that, and with data from the few studies that do exist for domestic livestock, we can possibly gain insights into what the non-lethal effects of predation risk on livestock might be.

Physiologically, we know that for humans, stress is a killer. Chronic, long-term stress can greatly increase the chances of dying from a variety of causes such as stroke and heart attack. It affects a person's attitude, energy levels, and even sexual performance. For most of us, this stress comes from non-life threatening factors, for example work, relationships, and so on. However, for the many people living under wartime conditions, soldiers to civilians, this stress is indeed related to life-threatening causes, which can persist long after the immediate threat is gone, for example post-traumatic stress disorder. These circumstances are indeed similar to those faced daily by prey and so it is not hard to assume that any level of predation risk can induce stress and its related effects on prey. And this assumption is backed up by research. As with humans, stress in animals produces higher levels of stress-related hormones in the blood stream, the three major ones being adrenaline, cortisol, and norepinephrine. Also, as with humans, the level of these hormones can easily be measured and used as indicators of stress levels.

Do animals exhibit higher levels of these stress hormones when under predation risk? For a variety of species, higher levels of stress hormones have been shown to be related to higher levels of predation risk (Boonstra et al., 1998; Faure et al., 2003; Sheriff et al., 2010). Furthermore, these higher levels of stress appear to have direct impacts on reproductive output (Creel et al., 2007; Sheriff et al., 2010) and survival (Creel et al., 2007). Stress from predation risk then, appears to have not only health consequences, but these effects translate to population effects that can affect the number of animals in an area beyond those killed by the predators.

Relative to fear-induced habitat changes, again, ample evidence exists of a variety of species adjusting how they use the landscape: their landscape of fear (Hernández and Laundré, 2005; Abu Baker and Brown, 2013; Nicholson et al., 2014). Under the risk of predation, animals are seeking out those habitats where they will either not encounter predators as much or if they do, have a better chance of escaping. As it turns out, predators are also responding to how habitat affects their hunting ability. Predators are selecting habitats where they are more lethal, ones being avoided by their prey for the same reason. As a result, contrary to what we would think, predators actually spend more time in areas that have less but more vulnerable prey (Laundré et al., 2009)! In this two-player game of cat and mouse, predator and prey seem to be spacing themselves out over the landscape.

Do these changes in habitat by the prey due to predation risk affect their foraging efficiency? It appears that it does. In one of the initial studies of this question, we demonstrated that elk under predation risk from wolves had lower quality diets than elk without wolves (Hernández and Laundré, 2005). More recent studies have also found further evidence that prey under predation risk will have a reduction in foraging efficiency (Christianson and Creel, 2010). Why does this happen? It appears that because prey animals are seeking out the safe areas, these areas tend to get overused and so nutritional quality of the areas declines. They then are faced with the choice of either going to areas of higher risk, which will have higher nutrition because of low use, or staying in the over used but safe areas.

Does the decrease in nutritional levels have health and population effects? Data again from population studies of elk and snowshoe hare under predation risk seem to indicate so (Creel et al., 2007; Sheriff et al., 2009).

Based on the data, it appears that the non-lethal impacts of predators can have subtle consequences for the health and survival of prey as they adjust to predation risk through fear. The question then becomes whether these effects can be seen in domestic animals raised extensively where there are predators. As mentioned, there have been few field studies of this. However, what we do know from a multitude of clinical and closed field trials is that domestic animals do respond to fear. Cattle are known to respond differently to gentle versus aggressive handlers (Rushen et al., 1999). They and many other domestic animals will show fear under a variety of stressful situations (Bouissou et al., 1996; Forkman et al., 2007; Richard et al., 2008; Wang et al., 2013; Zulkifli,

2013). This fear is also known to impact the productivity and welfare of these animals (Zulkifli, 2013). However, how about the wild? Do they, as their wild relatives, show fear of predators? And if so, will it also affect productivity and welfare?

Howery and DeLiberto (2004) asked the same questions in 2004 and using wildlife models, as I have, concluded that there was the potential, but at that point no field work existed. Since then, there have been a few studies specifically looking at non-lethal impacts of predators on livestock. One of the first studies was by Oakleaf et al. (2003) who looked at both lethal and non-lethal (movements) impacts of wolves on calves. Their results of the lethal effects mirrored what has already been discussed (total predation rate by wolves was 1.7%, cause-specific mortality rate was less than non-predation causes; 30.7% versus 61.5%). As for effects of wolves on the movements of calves, they found no difference between areas of high and low overlap of calves with known wolf movements.

Kluever et al. (2008) studied whether range cattle would increase their vigilance levels when faced with predation risk from wolves and cougars. They found that cattle had a relatively low level of vigilance (around 3% of their foraging time) compared to elk (12 to 30%) and bison (9.6 to 18.9%) (Laundré et al., 2001). This low level of vigilance may indicate a low level of perceived predation risk by cattle, as was seen with bison compared to elk (Laundré et al., 2001). However, as with elk, female cows with calves did have a higher level of vigilance (4.2% vs 2.0%) and females who had their calves preyed upon had even higher levels (up to 48%),

which decreased to the lower levels within three days. Overall, the conclusions of this study were that cattle were changing their vigilance levels when faced with higher predation risk (females with calves). Whether these increases in vigilance decreased the foraging efficiency of cattle to a significant level was not addressed in this study.

In 2010, Laporte et al. (2010) studied whether wolves would affect cattle and elk movement behaviors. Their results indicated that cattle, like elk, adjusted their movement behavior in the presence of predation risk from wolves. However, their study also did not assess if these changes affected the energy requirements or feeding efficiency of cattle.

Steele et al. (2013) attempted to model the potential economic impacts of three non-lethal factors (decreased weaning weights, decreased conception rates, and increased cattle sickness) on the overall cattle operation. They admitted they had no data other than interviews from diverse producers for the basis of their low to high effects. Their modeling efforts demonstrated a significant non-lethal economic effect. However, it is difficult to determine if these results are realistic or just artifacts of the input factors they used, which may or may not be realistic.

Ramler et al. (2014) were the first to try to quantify non-lethal effects of wolves on calf weights based on field data. Their study compared 18 ranches with differing levels of wolf home range overlap and confirmed wolf depredations. Of interest is that they found a positive correlation between wolf home range overlap and calf weights, in other words calves from ranches with greater wolf home range overlap weighed more.

Their explanation for this is that there were also confounding animal husbandry and environmental factors that might have contributed more to weight differences than the presence of wolves. They also observed that the ranches with wolf overlap also had higher levels of plant productivity resulting in higher calf weights, also making the areas attractive to wolves because of possibly higher native prey levels. However, they did not have data on native prey abundance. Ramler et al. (2014) did find a negative relationship between calf weights and whether a confirmed wolf predation event had occurred on the ranch. The estimated weight difference was approximately 22 pounds, which translated to an economic loss of around US$25 per head. Unfortunately, as they observed with wolf home range overlap, there are many confounding factors for which they did not have data that could have contributed to the weight differences. As they noted, animal husbandry differences can contribute to weight differences. Thus, if there was an association between predation incidences and level of animal husbandry, not all the weight differences could be attributed to the occurrence of predation on a ranch. As with wild ungulates (White et al., 2011), it can be difficult to separate predation effects on body condition from environmental variation and in this case, the added differences in animal husbandry.

In conclusion, it appears that domestic livestock also respond to risk with increased fear and stress and changes in behavior. Unfortunately, the question as to whether all this is of consequence to the survival and foraging efficiency of livestock on the range is not clear at this time. Though there are some indications,

it is also uncertain if changes in foraging efficiency affect productivity, for example weight gain, of range animals. Additionally, it is also unclear if fear of predation can be attributed to some of the non-predator-related deaths because of stress or decreases in nutritional quality. It is hoped that future research will help clarify the results obtained so far and help determine the importance of non-lethal impacts of predators on domestic livestock. Much more research is also needed to investigate the possible relationship of these non-lethal effects of the ecology of fear to actual economic loss to the extensive management operator.

It should be pointed out that we also don't know if the non-lethal control of predation discussed earlier might also reduce these non-lethal impacts of predation. It is possible that herds or flocks protected by guard dogs may not only be safer from being killed but also may experience lower fear and thus stress levels. Interestingly, from surveys, non-lethal controls of predation do seem to reduce the stress of ranchers who use guard animals (Green and Woodruff, 1983)! It seems that Virgil was right: "You'll *never fear* midnight thieves in the stables, attacks of wolves [my emphasis]."

AN ALTERNATIVE METHOD FOR DEALING WITH LIVESTOCK PREDATION

Although we have talked extensively about ways of reducing the already low level of predation on livestock in extensive management operations, a brief discussion is needed on an alternative to reducing predation: economic compensation

for losses. Under this system, it is recognized that having predators on the
landscape does incur economic losses
to livestock owners. However, instead
of trying to reduce these losses, usually by predator removal, owners are
compensated for the losses in return for
leaving the predators on the landscape.
The idea is that if we as a society consider having predators in an area to be
ecologically desirable, then we should
be willing to compensate these owners
for allowing predators to occur there.
The most commonly practiced compensation scheme is compensation for
actual losses of animals to predators
(Bruscino and Cleveland, 2004). This
compensation can come from govern-
mental agencies or private organizations. For example, many conservation
organizations, although predation losses
from wolves are indeed low, have set
up compensation funds to pay livestock
owners for the animals killed by wolves.
Whether private funds for specific predators, in this case wolves, or government
funds or credits for general predation,
the basic structure is "after the fact"
payments, referred to as ex-post compensation schemes (Schwerdtner and
Gruber, 2007). Under these plans the
livestock owner has to demonstrate that
the animal was killed by the predator
before compensation is given. For the
organization that is paying, this is considered a safeguard against abuse of the
system by the livestock owner claiming
deaths from other causes, which as we
saw is a majority of losses, as being from
predators. However, for the livestock
owners, this puts the burden of proof
on them and often deaths from predation can be difficult to verify. Added to
this, if the compensation is for a specific

predator, for example wolves, is the further difficulty of separating deaths from
different predators. There is also little
compensation for economic losses from
ancillary costs such as potential weight
loss or increased veterinary costs for the
animals that are not killed, and possibly
affected by the non-lethal impacts of
predators.

What has resulted is a system that
does not often satisfy either party and
tends to build distrust rather than the
cooperation that is needed for this
system to work (Treves et al., 2009). For
the payee, they often still feel that the
livestock owner is exaggerating losses.
For the owners, they feel that they are
being overburdened with compiling evidence and are often still denied legiti-
mate claims (Mattisson et al., 2011).

In most of Europe where livestock
owners also face predation from wolves
as well as several other large predators,
a similar ex-post compensation scheme
is used. However, in Sweden, an alternate system of compensation has developed. Instead of paying after a predation
event has occurred, agencies compensate herders for the potential loss of
animals. Based on past experience and
estimates of predator, native prey, and
livestock densities, estimates of how
many animals a header on average might
lose to predators per year are mutually
arrived at. The economic value of these
potential losses is estimated and paid
to the livestock owner up front. What
this system does is to remove the
points of contention so common in the
ex-post compensation system. There is
no burden of proof on the owners and,
once set, no overpaying for losses over
the estimate. This system assumes
predators will kill a certain number of

domestic livestock; it is part of the cost of doing business of raising livestock while maintaining predators on the landscape. Additionally, it actually encourages better animal husbandry as if a livestock owner actually has fewer animals killed as allotted, the predetermined amount is still received. Though far from perfect, this system does have advantages over ex-post compensation plans and could provide a model for other extensive management systems (Schwerdtner and Gruber, 2007, Mattisson et al., 2011). Such a system would also be conducive to compensation for non-lethal effects, for example potential weight losses or increased veterinary costs, in that once future research establishes what those costs are, they can be incorporated into the annual payment provided the livestock owner. There is not sufficient space here to discuss and compare these plans in detail. Swenson and André (205) and Schwerdtner and Gruber (2007) do provide some detailed comparisons and analyses of the two systems. These and future work should provide a basis for discussions of how to develop a mutually agreed method of compensation that could aid in attaining coexistence of predators with extensive livestock management practices.

LOOKING AHEAD

In ending, we need to look to the future regarding the welfare of livestock in extensive systems under the risk of predation and ask what challenges will need to be addressed. In doing so, we have to frame it in the dichotomy that exists in such systems. It is understood by most that extensively raised livestock owners

do have a right to protect their livestock from imminent attack by any predator. However it also has to be recognized that they have an ecological responsibility to maintain intact functional ecosystems that they depend on to raise their livestock and so much more. Meeting that ecological responsibility is one of the major challenges extensive livestock operations face today and in the future. Extensive livestock managers have to share the landscape with other "stakeholders" and users, often with more economic benefit than raising livestock. On public lands especially in the US, the use of the landscape for extensive management is heavily subsidized by the general public who use these lands and the wildlife living there for enjoyment. In return, extensive livestock managers have the responsibility of maintaining viable populations of all wildlife. This is especially true of predators that, based on the experiences in Yellowstone National Park, can be an economic engine for tourism (Duffield et al., 2008). Fortunately, as I have pointed out repeatedly in this chapter, predation is a fact but its severity has often been overemphasized. Except for individualized cases, the number of animals lost and the economic burden of these losses is low at an overall level. It even becomes less significant when compared to other causes of loss of livestock. As a result, the economics of predation never justify the blanket removal of any native predator from the landscape.

However, livestock do get killed by predators and the major responsibility of society as a whole is resolving these conflicts between predation and extensive management of livestock. The overall challenge, however, in resolving these conflicts should not be elimination of the predators but coexistence with them. Unfortunately, North American, and including South American, herders have too easily relied on extermination in the past and even now and have not developed alternative methods for coexistence. Fortunately, European and Asian herders who have been coexisting with predators for millennia have evolved various techniques to make this coexistence viable. These techniques can and have been implemented successfully in American extensive livestock management systems. However, as favorable as the outcome might be, use of these techniques is still low in the Americas and many livestock owners are reluctant to use them for a variety of reasons. As a result, one of the important challenges in looking ahead is how to implement more widespread use of non-lethal methods of controlling predation. Such implementation would be a major step in achieving coexistence with predators but still enable ranchers to use the land to raise livestock in extensive management systems.

Consequently, the biggest challenge facing modern extensive management of livestock is maintaining the viable ecosystems in which it is practiced. This includes maintaining ecologically viable populations of native predators. To meet that challenge we need more efforts on how that balance between economic viability and ecosystem integrity can be achieved. I hope that this chapter has provided some insights not into how we win the war against predators but into developing a truce that we and the native predators can both live with.

REFERENCES

Abu Baker, M.A., and Brown, J.S. (2013) Foraging and habitat use of common duikers, *Sylvicapra grimmia*, in a heterogeneous environment within the Soutpansberg, South Africa. *African Journal of Ecology* 52, 318–327.

Allen, L.R. (2014) Wild dog control impacts on calf wastage in extensive beef cattle enterprises. *Animal Production Science.* 54, 214–220.

Allen, L.R., and Sparkes, E.C. (2001) The effect of dingo control on sheep and beef cattle in Queensland. *Journal of Applied Ecology* 38, 76–87.

Amit, R., Gordillo-Chávez, E.J., and Bone, R. (2013) Jaguar and puma attacks on livestock in Costa Rica. *Human-Wildlife Interactions*, 7, 77–84.

Andelt, W.F. (1992) Effectiveness of livestock guarding dogs for reducing predation on domestic sheep. *Wildlife Society Bulletin* 20, 55–62.

Andelt, W.F. (2004) Use of livestock guardian animals to reduce predation on sheep and livestock. *Sheep & Goat Research Journal* 19, 72–75.

Andelt, W.F., and Hopper, S.N. (2000) Livestock guard dogs reduce predation on domestic sheep in Colorado. *Journal of Range Management* 53, 259–267.

Anonymous (2014) A history of wildlife in North America – MarineBio.org. MarineBio Conservation Society. Web. Available at: www.marinebio.org/oceans/conservation/moyle/ch2-2/

Arjo, W.M., and Peltscher, D.H. (2004) Coyote and wolf habitat use in northwestern Montana. *Northwest Science* 78, 24–32.

Banerjee, K., Jhala, Y.V., Chauhan, K.S., and Dave, C.V. (2013) Living with lions: The economics of coexistence in the Gir Forests, India. PLOS ONE 8, 1–11: e49457, doi: 10.1371/journal-pone.0049457

Binkley, D., Moore, M.M., Romme, W.H., and Brown, P.M. (2006) Was Aldo Leopold right about the Kaibab deer herd? *Ecosystems* 9, 227–241.

Boonstra, R., Hik, D., Singleton, G.R., and Tinnikov, A. (1998) The impact of predation-induced stress on the snowshoe hare cycle. *Ecology Monographs* 68, 371–394.

Bouissou, M.F., Porter, R.H., Boyle, L., and Ferreira, G. (1996) Influence of a conspecific image of own vs. different breed on fear reactions of ewes. *Behavioural Processes* 38, 37–44.

Brown, J.S., Laundré, J.W., and Gurung, M. (1999) The ecology of fear: Optimal foraging, game theory, and trophic interactions. *Journal of Mammalogy* 80, 385–399.

Bruscino, M.T., and Cleveland, T.L. (2004) Compensation programs in Wyoming for livestock depredation by large carnivores. *Sheep & Goat Research Journal* 19, 47–49.

Bueno-Cabrera, A., Hernández-Garcia, L., Laundré, J., Contreras-Hernandez, A., and Shaw, H. (2005) Cougar impact on livestock ranches in the Santa Elena Canyon, Chihuahua, Mexico. *Mountain Lion Workshop* 8, 141–149.

Christianson, D., and Creel, S. (2010) A nutritionally mediated risk effect of wolves on elk. *Ecology* 91, 1184–1191.

Conforti, V.A., and Cascelli de Azevedo, F.C. (2003) Local perceptions of jaguars (*Panthera onca*) and pumas (*Puma concolor*) in the Iguaçu National Park area, south Brazil. *Biological Conservation* 111, 215–221.

Coppinger, L., and R. Coppinger (2007) Dogs for herding and guarding livestock. In: Grandin, T. (Ed.) *Livestock Handling and Transport*, 3rd edn. CAB International, Oxford, UK. pp. 199–213.

Coppinger, R., Coppinger, L., Langeloh, G. L., Gettler, L., and Lorenz, J. (1988) A decade of use of livestock guarding dogs. *Proceedings of the Thirteenth Vertebrate Pest Conference* 13, 209–214. Available at: http://digitalcommons.unl.edu/vpcthirteen (accessed 6 November 2014).

Creel, S., Christianson, D., Liley, S., and Winnie Jr. J.A. (2007) Predation risk affects reproductive physiology and demography of elk. *Science* 315, 960.

Dalponte, J.C. (2002) Jaguar diet and predation on livestock in the northern Pantanal, Brazil. In: Medellín, R.A. et al. (Eds.) *El jaguar en el nuevo milenio. Una evaluacion de su estado, deteccion de prioridades y recomendaciones para la conservacion de los jaguars en America*. FCE, UNAM, Wildlife Conservation Society.

Downward, R.J., and Bromell, J.E. (1990) The development of a policy for the management of dingo populations in South Australia. *Proceedings of the 14th Vertebrate Pest Conference* (L.R. Davis and R.E. March, Eds). University of California, Davis. pp 241–244.

Duffield, J.W., Neher, C.J., and Patterson, D.A. (2008) Wolf recovery in Yellowstone: Park visitor attitudes, expenditures, and economic impacts. *Yellowstone Science* 16, 21–25.

Fagerstone, K.A., Johnston, J.J., and Savarie, P.J. (2004) Predacides for canid predation management. *Sheep & Goat Research Journal* 19, 76–79.

Faure, J.M., Val-Laillet, D., Guy, G., Bernadet, M.D., and Guémené, D. (2003) Fear and stress reactions in two species of duck and their hybrid. *Hormones and Behavior* 43, 568–572.

Fleming, P., Corbett, L., Harden, R., and Thomson, P. (2001) *Managing the impacts of dingos and other wild dogs*. Bureau of Rural Sciences, Canberra.

Fleming, P.J.S., and Korn, T.J. (1989) Predation of livestock by wild dogs in Eastern New South Wales. *Australian Rangeland Journal* 11, 61–66.

Forkman, B., Boissy, A., Meunier-Salaün, M.C., Canali, E., and Jones, R.B. (2007) A critical review of fear tests used on cattle, pigs, sheep, poultry and horses. *Physiology & Behavior* 92: 340–374.

Gazzola, A., Capitani, C., Mattioli, L., and Apollonio, M. (2008) Livestock damage and wolf presence. *Journal of Zoology* 274, 261–269.

Gehring, T.M., VerCauteren, K.C., and Landry J.M. (2010) Livestock protection dogs in the 21st century: Is an ancient tool relevant to modern conservation challenges? *BioScience* 60, 299–308.

Gese, E. (2005) Demographic and spatial responses of coyotes to changes in food and exploitation. *Proceedings of 11th Wildlife Damage Management Conference* 11, 271–285 Available at: http://digital-commons.unl.edu/icwdm_wdmconfproc (Accessed 6 November 2014).

Green, J.S., and Woodruff, R.A. (1983) The use of three breeds of dogs to protect rangeland sheep from predators. *Applied Animal Ethology* 11, 141–161.

Green, J.S., and Woodruff, R.A. (1984) Livestock-guarding dogs for predator control: Costs, benefits, and practicality. *Wildlife Society Bulletin* 12, 44–50.

Green, J. S., and Woodruff, R. A. (1993) Livestock guarding dogs: Protecting sheep from predators. Available at: http://hdl.handle.net/2027/umn.31951d012181083 (accessed 6 November 2014).

Hernández, H., and Laundré, J.W. (2005) Foraging in the 'landscape of fear' and its implications for the habitat use and diet quality of elk *Cervus elaphus* and bison *Bison bison*. *Wildlife Biology* 11, 215–220.

Hewitt, L. (2009) *Major economic costs associated with wild dogs in the Queensland grazing industry*. Blueprint for the Bush. Queensland, Australia.

Holmes, B.R., and Laundré, J.W. (2006) Use of open, edge and forest areas by pumas *Puma concolor* in winter: Are pumas foraging optimally? *Wildlife Biology* 12, 201–209.

Hoogesteijn, R., Hoogesteijn, A., and Mondolfi, E. (1993) Jaguar predation and conservation: Cattle mortality caused by felines on three ranches in the Venezuelan Llanos. *Symposium of the Zoological Society of London* 65, 391–407.

Horsley, S.B., Stout, S.L., and DeCalesta, D.S. (2003) White-tailed deer impact on the vegetation dynamics of a northern hardwood forest. *Ecological Applications* 13, 98–118.

Hosseini-Zavarei, F., Farhadinia, M.S., Beheshti-Zavareh, M., and Abdoli, A. (2013) Predation by grey wolf on wild ungulates and livestock in central Iran. *Journal of Zoology* 290, 127–134.

Howery, L.D., and DeLiberto, T.J. (2004) Indirect effects of carnivores on livestock foraging behavior and production. *Sheep & Goat Research Journal* 19, 53–57.

Huntsinger, L., and Starrs, P.F. (2006) Grazing in arid North America: A biogeographical approach. *Sécheresse* 17, 219–233.

Iliopoulos, Y., Sgardelis, S., Koutis, V., and Savaris, D. (2009) Wolf depredation on livestock in central Greece. *Acta Theriologica* 54, 11–22.

Joly, D.O., Heisey, D.M., Samuel, M.D., Ribic, C.A., Thomas, N.J, Wright, S.D., and Wright, I.E. (2009) Estimating cause-specific mortality rates using recovered carcasses. *Journal of Wildlife Diseases* 45, 122–127.

Kline, A.S. (2002) Translation of Virgil: Georgics Book III Livestock farming. Available at: www.poetryintranslation.com/PITBR/Latin/VirgilGeorgicsIII.htm#_Toc534252746 (Accessed 6 November 2014).

Kluever, B.M., Breck, S.W., Howery, L.D., Krausman, P.R., and Bergman, D.L. (2008) Vigilance in cattle: The influence of predation, social interactions, and environmental factors. *Rangeland Ecology & Management* 61, 321–328.

Kolowski, J.M., and Holekamp, K.E. (2006) Spatial, temporal, and physical characteristics of livestock depredations by large carnivores along a Kenyan reserve border. *Biological Conservation* 128, 529–541.

Kuzyk, G.W., Kneteman, J., and Schmiegelow, F.K.A. (2004) Winter habitat use by wolves, *Canis lupus*, in relation to forest harvesting in west-central Alberta. *Canadian Field Naturalist* 118, 368–375.

Laporte, I., Muhly, T.B., Pitt, J.A., Alexander, M., and Musiani, M. (2010) Effects of wolves on elk and cattle behaviors: Implications for livestock production and wolf conservation. PLos ONE 5(8): e11954. Doi:10.1371/journal.pone.0011954.

Laundré, J.W. (2012) *Phantoms of the prairie: The return of cougars to the Midwest.* University of Wisconsin Press, Madison, Wisconsin.

Laundré, J.W., and Hernández, L. (2003) Winter hunting habitat of pumas *Puma concolor* in northwestern Utah and southern Idaho, USA. *Wildlife Biology* 9, 123–129.

Laundré, J.W., Hernández, L., and Altendorf, K.B. (2001) Wolves, elk, and bison: Reestablishing the "landscape of fear" in Yellowstone National Park, U.S.A. *Canadian Journal of Zoology* 79, 1401–1409.

Laundré, J.W., Calderas, J.M.M., and Hernández, L. (2009) Foraging in the landscape of fear, the predator's dilemma: Where should I hunt? *The Open Ecology Journal* 2, 1–6.

Laundré, J.W., Hernández, L., and Ripple, W.J. (2010) The landscape of fear: Ecological implications of being afraid. *The Open Ecology Journal* 3, 1–7.

Leopold, A. (1943) Deer irruptions. *Wisconsin Conservation Bulletin* 8, 1–11.

Leopold, A., Sowls, L.K., and Spencer, D.L. (1947) A survey of over-populated deer ranges in the United States. *Journal of Wildlife Management* 11, 162–177.

Li, X, Buzzard, P., Chen, Y., and Jiang, X. (2013) Patterns of livestock predation by carnivores: Human-wildlife conflict in Northwest Yunnan, China. *Environmental Management* 52, 1334–1340.

Mastro, L.L. (2011) Life history and ecology of coyotes in the mid-Atlantic states: A summary of the scientific literature. *Southeastern Naturalist* 10, 721–730.

Mattisson, J., Odden, J., Nilsen, E.B., Linnell, J.D.C., Persson, J., and Andrén, H. (2011) Factors affecting Eurasian lynx kill rates on semi-domestic reindeer in northern Scandinavia: Can ecological research contribute to the development of a fair compensation system? *Biological Conservation* 144, 3009–3017.

Mazzolli, M., Graipel, M.E., and Dunstone, N. (2002) Mountain lion depredation in southern Brazil. *Biological Conservation* 105, 43–51.

Mech, L.D., Smith, D.W., Murphy, K.M., and MacNulty, D.R. (2001) Winter severity and wolf predation on a formally wolf-free elk herd. *Journal of Wildlife Management* 65, 998–1003.

Michalski, F., Boulhosa, R.L. P., Faria, A., and Peres, C.A. (2006) Human-wildlife conflicts in a fragmented Amazonian forest landscape: determinants of large felid depredation on livestock. *Animal Conservation* 9, 179–188.

Namgail, T., Fox, J.L., and Veer Bhatnagar, Y. (2007) Carnivore-caused livestock mortality in Trans-Himalaya. *Environmental Management* 39, 490–496.

Nellis, C.H., and Keith, L.B. (1968) Hunting activities and success of lynxes in Alberta. *Journal of Wildlife Management* 32, 718–722.

Nelson, M.P., Vucetich, J.A., Peterson, R.O., and Vucetich, L.M. (2010) The Isle Royale wolf-moose project (1958–present) and the wonder of long-term ecological research. *Endeavour* 35, 30–38.

Nicholson, K.L., Milleret, C., Månsson, J. (2014) Testing the risk of predation hypothesis: The influence of recolonizing wolves on habitat use by moose. *Oecologia* 176, 69–80.

Nowak, S., Myslajek, R.W., and Jedrzejewska, B. (2005) Patterns of wolf *Canis lupus* predation on wild and domestic ungulates in the Western Carpathian Mountains (S. Poland). *Acta Theriologica* 50, 263–276.

Oakleaf, J.K., Mack, C., and Murray, D.L. (2003) Effects of wolves on livestock calf survival and movements in central Idaho. *Journal of Wildlife Management* 67, 299–306.

Palmeira, F.B.L., Crawshaw, P.G. Jr., Haddad, C. M., Ferraz, K.M.P.M.B., and Verdade, L.M. (2008) Cattle depredation by puma (*Puma concolor*) and jaguar (*Panthera onca*) in central-western Brazil. *Biological Conservation* 141, 118–125.

Patterson, B.D., Kasiki, S.M., Selempo, E., and Kays, R.W. (2004) Livestock predation by lions (Panthera leo) and other carnivores on ranches neighboring Tsavo National Parks, Kenya. *Biological Conservation* 119, 507–516.

Peebles, K.A., Wielgus, R.B., Maletzke, B.T., and Swanson, M.E. (2013) Effects of remedial sport hunting on cougar complaints and livestock depredations. PLos ONE 8(11): e79713. Doi:101371/journal. pone.0079713.

Penna, A.N. (1999) *Nature's bounty: Historical and modern environmental perspectives.* M.E. Sharpe, Inc. Armonk, New York.

Pierotti, R. (2011) *Indigenous knowledge, ecology, and evolutionary biology.* Routledge, New York, New York.

Ramler, J.P., Hebblewhite, M., Kellenberg, D., and Sime, C. (2014) Crying Wolf? A spatial analysis of wolf location and depredations on calf weight. *American Journal of Agricultural Economics* 96, 631–656.

Richard, S., Wacrenier-Ceré, N., Hazard, D., Saint-Dizier, H., Arnould, C., and Faure, J.M. (2008) Behavioural and endocrine fear responses in Japanese quail upon presentation of a novel object in the home cage. *Behavioural Processes* 77, 313–319.

Rigg, R., Findo, S., Wechselberger, M., Gorman, M.L., Sillero-Zubiri, C., and MacDonald, D.W. (2011) Mitigating carnivore-livestock conflicts in Europe: Lessons from Slovakia. *Oryx* 45, 272–280.

Rushen, J., Taylor, A.A., and de Passillé, A.M. (1999) Domestic animals' fear of humans and its effect on their welfare. *Applied Animal Behaviour Science* 65, 285–303.

Rust, N.A., Whitehouse-Tedd, K.M., and MacMillan, D.C. (2013) Perceived efficacy of livestock-guarding dogs in South Africa:

Implications for cheetah conservation. *Wildlife Society Bulletin* 37, 690–697.

Schwerdtner, K., and Gruber, B. (2007) A conceptual framework for damage compensation schemes. *Biological Conservation* 134, 354–360.

Sheriff, M.J., Krebs, C.J., and Boonstra, R. (2009) The sensitive hare: Sublethal effects of predator stress on reproduction in snowshoe hares. *Journal of Animal Ecology* 78, 1249–1258.

Sheriff, M.J., Krebs, C.J., and Boonstra, R. (2010) The ghosts of predators past: Population cycles and the role of maternal programming under fluctuating predation risk. *Ecology* 91, 2983–2994.

Smith, M.E., Linnel J.D.C., Odden, J., and Swenson, J.E. (2000) Review of methods to reduce livestock depredation: I Guardian animals. *Acta Agricuture Scandinavica* 50, 279–290.

Smits, D.D. (1994) The frontier army and the destruction of the buffalo: 1865–1883. *The Western Historical Quarterly* 25, 312–338.

Stahl, P., Vandel, J.M., Herrenschmidt, V., and Migot, P. (2001) Predation on livestock by an expanding reintroduced lynx population: Long-term and spatial variability. *Journal of Applied Ecology* 38, 674–687.

Steele, J.R., Rashford, B.S., Foulke, T.K., Tanaka, J.A., and Taylor, D.T. (2013) Wolf (*Canus lupus*) predation impacts on livestock production: Direct effects, indirect effects, and implications for compensation ratios. *Rangeland Ecology & Management* 66, 539–544.

Swenson, J.E., and Andrén, H. (2005) A tale of two countries: Large carnivore depredation and compensation schemes in Sweden and Norway. In: Woodroffe, R., Thirgood, S., and Rabinowitz, A. (Eds.) *People and wildlife–Conflict or coexistence?* Cambridge University Press, England. pp. 323–339.

Tchir, T.L., Johnson, E., and Nkemdirim, L. (2014) Deforestation in North America: Past, present and future. Regional Sustainability Development Review: Canada and USA. Vol I.

Tishler, W., and Witmer, C.S. (1986) The housebarns of east-central Wisconsin. *Perspectives in Vernacular Architecture* 2, 102–110.

Treves, A., Jurewicz, R.L., Haughton-Treves, L., and Wilcove, D.S. (2009) The price of tolerance: Wolf damage payments after recovery. *Biodiversity Conservation* 18, 4003–4021.

Tumenta, P.N., De Iongh, H.H., Funston, P.J., and Udo de Haes H.A. (2013) Livestock depredation and mitigation methods practiced by resident and nomadic pastoralists around Waza National Park, Cameron. *Oryx* 47, 237–242.

Urbigkit, C., and Urbigkit, J. (2010) A review: The use of livestock protection dogs in association with large carnivores in the Rocky Mountains. *Sheep & Goat Research Journal* 25, 1–8.

Van Bommel, L., and Johnson, C.N. (2012) Good dog! Using livestock guardian dogs to protect livestock from predators in Australia's extensive grazing systems. *Wildlife Research* 39, 220–229.

Wagner, C., Holzapfel, M., Kluth, G., Reinhardt, I., and Ansorge, H. (2012) Wolf (*Canis lupus*) feeding habits during the first eight years of its occurrence in Germany. *Mammalian Biology* 77, 196–203.

Wang, S., Hi, Y., Guo, F., Fu, W., Grossmann, R., and Zhao, R. (2013) Effect of corticosterone on growth and welfare of broiler chickens showing long and short tonic immobility. *Comparative Biochemistry and Physiology, Part A* 164, 537–543.

Wegge, P., Shrestha, R., and Flagstad, Ø. (2012) Snow leopard *Panthera uncia* predation on livestock and wild prey in a mountain valley in northern Nepal: Implications for conservation management. *Wildlife Biology* 18, 131–141.

Whitaker, J.O. Jr., and Hamilton, W.J. Jr. (1998) *Mammals of the Eastern United States*. Cornell University Press. Ithaca, New York.

White, P.J., Garrott, R.A., Hamlin, K.L., Cook, R.C., Cook, J.G., and Cunningham, J.A. (2011) Body condition and pregnancy in northern Yellowstone elk: Evidence for predation risk effects? *Ecological Applications* 21, 3–8.

Woodroffe, R., Lindsey, P., Romañach, S., Stein, A., and ole Ranah, S.M.K. (2005) Livestock predation by endangered African wild dogs (*Lycaon pictus*) in northern Kenya. *Biological Conservation* 124, 225–234.

Zeder, M.A. (2008) Domestication and early agriculture in the Mediterranean Basin: Origins, diffusion, and impact. *Proceedings of the National Academy of Sciences* 105, 11597–11604.

Zimmermann, B., Wabakken, P., and Dötterer (2003) Brown bear–livestock conflicts in a bear conservation zone in Norway: Are cattle a good alternative to sheep? *Ursus* 14, 72–83.

Zulkifli, I. (2013) Review of human-animal interactions and their impact on animal productivity and welfare. *Journal of Animal Sciences and Biotechnology* 4, 25–31.

chapter seven

The health of livestock in extensive systems

Pete Goddard

INTRODUCTION

In this volume about livestock welfare in extensive systems it is important to consider animal health, because health and welfare are inextricably linked and extensive systems can pose different or additional health challenges to those experienced by livestock in more intensive systems. These challenges may be due to difficulties with diagnosis and treatment as a result of the way animals are managed, different patterns of exposure to infectious agents, or health-related environmental aspects. As a result of these challenges, livestock may be more or less liable to experience disease, depending on the pathogenic organism concerned and the capacity, skill, and resources of the stockperson. This chapter will examine in a broad way the position of extensive livestock production systems regarding animal health and disease. Extensive livestock systems are so varied in location, structure, management practice, and many other features that it is not possible to provide a simple unitary guide to the health situation and disease concerns that might arise. Thus this chapter will sketch out some of the broad issues rather than provide a definitive guide to current and best practice around the world. Because in some extensive systems there is a very intimate relationship between the stockperson (and sometimes their family members) and their livestock, it is also necessary to consider whether such systems have a likelihood of altered zoonotic disease risk. This chapter will provide a small number of illustrative examples of specific diseases of interest and how they have been and are being tackled. Given, in some cases, the zoonotic disease risk may be considerable, it is appropriate to reflect on how the emerging One Health[1]

The James Hutton Institute, Aberdeen, UK

[1] One Health is a relatively new but increasingly adopted consideration of the interdependence of human, animal, and environmental health. The foundation of the One Health paradigm is multidisciplinary collaboration between a wide range of parties, encompassing many disciplines that impact human and animal health, food security, and food safety. As defined by the AVMA as "One Health is the collaborative effort of multiple disciplines – working locally, nationally, and globally – to attain optimal health for people, animals and our environment." AVMA, 2008.

approach has a bearing in terms of extensive livestock health. Even though land availability will be a limiting factor on extensive livestock numbers in the future, extensive systems nonetheless represent a major production sector and the mainstay of many rural communities.

It should be remembered that while many extensive systems have the allure of "naturalness" and the associated perception of welfare benefit and thus credentials important to ethical consumers, many such systems cannot necessarily deliver a high level of individual health care as will be discussed; however, this ethical dilemma per se will not be considered further in this chapter. There are parallel questions to resolve as to whether extensive livestock systems are better or worse than intensive systems in terms of their overall environmental footprint.

It is not the intention of this chapter to provide an extensive description of the wide range of diseases which can affect extensively managed livestock. In particular, many extensive systems are found in tropical countries but this short chapter will not be a treatise on tropical animal diseases which can affect animals in these areas, whatever the system. It will, nevertheless, acknowledge the challenge for everyone involved in food production to maintain healthy animals in order to enhance food security and do so in a sustainable way.

DESCRIPTION OF SYSTEMS AND SPECIES

The question – "what constitutes an extensive livestock system?" has been explored in detail elsewhere in this book but is worth re-framing in the context of animal health and disease. For the purpose of this chapter, I will simply reiterate that such systems cover herbivores grazing and browsing rangelands, forestlands, mountain regions, grasslands, and open pastures. In this way, extensive systems include not only animals making a living from landscapes offering indigenous vegetation but also from landscapes providing introduced plant species. Such systems are usually ones in which animals have little contact with humans, including infrequent handling (Hodgson, 1990). However, a little licence will be used in order to capture animal health issues across the wide range of commonly-found systems. For example, many extensive systems rely on continual movement of livestock over large areas covered by sparse or poorly nutritious vegetation and sometimes in these systems livestock is tended by a single stockperson who may be in close contact with a large number of animals. In some cases these systems are described as marginal and practised by subsistence farmers or subsistence communities though here we may see resource-efficient integrated systems using agricultural by-products (particularly for small ruminants) as part of the livestock production cycle. Such systems are usually good for soil fertility and can help rejuvenate significant areas of land. Extensive systems also cover nomadic herdsmen and seasonal transhumance where stockpersons move with their cattle, sheep, or goats to live with them to take advantage of summer or seasonal grazing. In order to distinguish all these from the alternatives, I need a collective term for the latter but "conventional" is not appropriate – certainly from a truly historical perspective; throughout the remainder of this chapter I will use "non-extensive" as the collective descriptor to allow a distinction to be drawn when required.

Which Species are Under Consideration?

Extensive livestock systems are mostly based on ruminant herbivores. Their significant advantages over monogastric animals are largely due to the ability of the rumen to act as a buffer against fluctuating feed and water availability. Ruminants can also adapt and become tolerant to heat (or cold). Thus, goats, sheep, and cattle of a number of breeds are commonly found. To a lesser extent, species of deer (including reindeer), buffaloes, camels, and bison are also managed extensively.

Where in the World?

Extensive systems are found in all areas of the world from the Scottish uplands, the central Asian mountains to the Australian outback. The arid and semi-arid regions of the world encompass around 40 million pastoral graziers. A large number of people engaged in seasonal transhumance (Squires and Sidahmed, 1997). Many extensive systems are therefore found in cold, high-altitude mountainous, and upland regions of the world where cropping is not feasible. The physical terrain has an impact on animal health; this can be either as a result of the physical risks, the altered risk of exposure to disease-causing agents, or the capability of livestock keepers and veterinarians to deliver diagnostic, preventative, or treatment programmes. These areas also cover the breadth of eco-climatic regions

and expose livestock to potentially challenging conditions that may impact on their general fitness and, consequently, disease susceptibility, either on a continuous basis or at particular times of the year. Also, fluctuating environmental conditions have a big impact on arthropod disease vectors leading to seasonal outbreaks of disease in livestock. As most extensive livestock are not housed, their exposure to arthropod vectors cannot be controlled easily. Many livestock keepers have learnt to adapt grazing practices to minimise the likelihood of vector-borne disease but abnormal weather conditions and the increasing influence of climate change may overshadow traditional practices.

Some examples of extensive systems range from the nomadic or transhumance/migratory, to sheep or cattle ranching or rangeland systems practised over vast areas, in Australia for example. Nomadic herdsmen are always on the move and have no fixed home (though they may remain in some regions for longer periods of time). These herdsmen will keep a regular close watch on their animals to monitor their condition and spot any disease conditions. Nomads interact with pastoral communities they meet along the way and may be able to provide information updates (including about health) and health services to these sedentary pastoralists. Traditional transhumance systems, where stock are moved to alternative grazing for part of the year can pose a challenge when regional or national borders or fence lines are also used to enforce disease control strategies. This can be a problem where, for example, transhumance was practised to move animals away from seasonal tsetse fly areas for part of the year.

There is a considerable current and urgent debate about the contribution of grazing animals to climate change through greenhouse gas (GHG) emissions. There are various estimates but livestock are considered to contribute about 15% of human-induced GHG emissions. Most extensive systems rely on grazing and browsing indigenous (usually) vegetation but there is grazing in many non-extensive systems too. However, in extensive systems, livestock are generally slower to reach market weight and thus are GHG producers for a longer period. Yet set against this, the often lower replacement rate for breeding or milking females is beneficial; these and numerous additional factors make calculating the climate change footprint anything other than simple. In many cases, livestock in extensive systems are able to survive in areas and on resources that would otherwise not be utilized. One way to mitigate the overall impact of livestock husbandry in terms of the GHG equation for a system overall, is to ensure that as many as possible of the young born reach maturity, and that animals remain healthy and reach the market in optimal condition. Thus, preventive medicine programmes and effective treatment regimens are especially important in extensive systems where animals can take a long time to reach market condition and, equally, ensuring good health reduces the female replacement rate. With education and training of stockpersons and, in some cases, financial support, production could be enhanced. There is gathering momentum for farmers to do more to adopt systems which do not damage the environment and ensuring good health is part of this responsibility.

THE BIG ISSUES

Livestock managed in extensive systems are not fundamentally any different to those managed in more intensive ways when it comes to evaluating disease risk. The same epidemiological principles apply though the parameters which define disease susceptibility, occurrence, and resolution will vary. The challenge for disease control as in any system can be summarised by "identification and action." Essentially the most significant ways that extensive systems differ from more non-extensively managed livestock relate to the inability (in many cases) to rapidly identify and diagnose disease (including local, regional, and national surveillance to detect changes in the likely level of threat). There are then more immediately logistical problems related to treatment, mostly as a result of the difficulty or impossibility to gather all at-risk animals and contacts for effective treatment and control measures including quarantine to be applied. Also, reservoirs of disease in the local wildlife community can hamper disease control efforts. Other issues to take into account are the relative risk of current exposure and the relation of this to previous exposure/immunity.

Principally in relation to stockpersons, their actions relate to their knowledge, motivation, and capability, the latter being influenced by facilities available. For example, in nomadic systems fixed facilities, even when available, can be very limited. External support in the widest sense may also be lacking. For example, options for training may be poor in remote areas (or where stockpersons need to travel long distances to a central delivery point) and many extensive flocks or herds are often in the care of single people, making it difficult for them to take time out for training to increase their skills. Part of this difficulty can arise through failure of contemporary awareness and knowledge transfer rather than the lack of appetite for this from the stockpersons themselves (though recently smartphone systems and on-line learning opportunities have evolved in a number of areas to improve knowledge exchange).

Mitigation of these big issues can be in the form of preventive strategies such as vaccination or the use of disease-resistant stock. However, even when vaccines are available, for various reasons they are often not widely used. In some areas, *Bos indicus* genotypes of cattle are preferred as they are more resistant to ticks and helminth parasites and can have a greater resistance to protozoa such as some *Babesia spp.* (Bock et al., 1997). Grazing management (for example by avoiding certain areas at times of the year when arthropod vector activity is greatest) or surveillance of sentinel animals (if herd or flock preventative strategies can realistically be adopted) are important areas to consider. (Of course there is limited value in adopting surveillance if the results cannot be acted upon.)

If disease does occur, one course of action may be the use of antimicrobial or antiparasitic therapy. In the world-wide campaign to limit the emergence of antimicrobial resistance (AMR), there is a particular challenge to avoid blanket antibiotic treatment in rangeland situations. Here, because gathering animals can be very difficult, stockpersons may feel motivated for pragmatic reasons to treat all animals as there may be only infrequent occasions when whole herd or flock approaches can be applied when disease is present, anticipated, or not yet apparent. A recent analysis of surveillance for AMR is given in the 2014 report from the World Health Organisation (WHO, 2014). Here it is made clear that AMR within a wide range of infectious agents is a growing public health threat but there are important gaps in surveillance of AMR from food-producing animals; animals in extensive systems prove a particular surveillance challenge.

There are likely to be biosecurity concerns in extensive systems where many herds/flocks share common grazing or resources such as watering points. At these latter points in reality we can see a very intensive congregation of animals. This leads to the need for stockpersons or farmers' groups to cooperate in order to apply specific control measures within a defined treatment window based on the active period of the effect of the treatment itself. An example from the UK is in the attempts to control infections with the sheep scab mite (*Psoroptes ovis*). Spread is chiefly following sheep-to-sheep contact but mites can survive in the environment, in buildings, fences, or on livestock vehicles so direct contact is not always needed. One challenge under extensive conditions is to ensure that all susceptible animals in an epidemiological unit are gathered and treated, as infected individuals that escape treatment for whatever reason serve to reintroduce the disease after the treatment has worn off.

As a final point, moving livestock from extensive systems to non-extensive systems for marketing, breeding, or finishing may expose them to diseases to which they are naive and there are clear health risks through mixing during yarding and transport.

Some of these big issues will be

examined in more detail in the next section.

WHAT FACTORS INFLUENCE LIVESTOCK DISEASE RISK IN EXTENSIVE SYSTEMS?

Numerous factors can influence disease risk in any production system, including those considered extensive. This section will focus on some of the more significant risk factors that are at play and impact on extensive livestock specifically (Table 7.1.) These factors may either increase or decrease risk. Although the separation is not a strict one, in the first part of the table, the factors could be considered as potential hazards. In the second part of the table are factors over which the livestock keeper has some control and which may result in a greater or lesser risk of a problem occurring, sometimes through directly reducing the impact of the hazard. In addition, many extensive systems are found in areas where there is political unrest. This may, for example, impact on disease control programmes and the free movement of livestock.

Hazards Affecting Health ands Disease

Many extensive systems are found in the tropics. Despite well-developed mechanisms of thermoregulation, ruminants do

Table 7.1   Main factors influencing extensive livestock disease risk

---

HAZARDS
Eco-climatic conditions
Periodic food and water limitation or contamination of scarce resources
Exposure to native plant species and possible poisonous plants (See Chapter 5)
Deficiency diseases or toxicities
Exposure to arthropod vectors of disease and the ease of their control
Exposure to wildlife species which may be reservoirs of disease
Exposure to predators
Physical injury due to terrain
Natural disasters
Civil unrest/war

FACTORS THAT INCREASE/DECREASE RISK OF HAZARDS
Livestock species and breed/strain
Stocking density/animal contact
Large number of animals in one epidemiological unit
Frequency and intensity of inspection and interaction with stockperson
Rapidity and ease of diagnosing and treating affected and at-risk individuals
Having up-to-date information
Ability to significantly deliver biosecurity on rangelands (mixing of different groups of stock)
Ability to deliver whole herd/flock preventive measures including national/international disease control programmes
Ability to deliver severe weather protection or response to natural disasters
Parasite prevalence in relation to stocking density

---

not maintain strict homeothermy when under heat stress (Silanikove, 2000). Animals suffering from heat stress have a generally reduced biological function, including a decline in immune function. Heat stress can, for example, adversely affect the health of dairy cows (Collins and Weiner, 1968). Heat stress can be exacerbated if animals are driven or mustered under very hot conditions and this, it has been suggested, may increase lamb losses due to pneumonia (Black et al., 2005). Water deprivation may exacerbate heat stress and so be an important compounding factor under tropical conditions (especially if animals are pregnant or lactating) – more detail is given in Chapter 3. Heat is a major constraint on animal production in tropical and arid areas (Silanikove, 1992). Although some animals can cope with water access at infrequent intervals (probably once every two to three days for adapted, tolerant cattle, slightly longer for sheep and goats; Gregory, 2007) water shortage and the resultant intermittent large intake can cause significant detrimental physiological effects.

Poor health in extensive systems may be linked to an overall inadequate level of nutrition, not only in terms of macronutrients but also minerals and vitamins. These deficiencies can occur over both short and long time frames. Weak animals are also more prone to accidents and injury. It is not the intention to focus in a general way on nutrition in this chapter but there are recognised deficiency diseases which are specifically related via the plant to soil nutrient availability. Extensive livestock are more likely to be exposed to food and water shortages than in non-extensive systems. These can lead to reduced body condition and thence increased susceptibility to infectious disease and the effects of parasites. Droughts or floods which reduce available grazing can lead to animals in poorer body condition or starvation which leads to a reduction in the effectiveness of the immune system. Provision of supplementary vitamins and minerals can be of considerable help under these circumstances. Flooding events also increase the risk of exposure to diseases such as anthrax if the carcasses of buried animals which died of this disease are exposed and there may be a dramatic increase in the number of arthropod disease vectors which thrive in new swampy conditions; for example after extreme rainfall, the mosquito vectors of Rift Valley fever can increase dramatically.

Plant toxicities are dealt with elsewhere in this book: please see Chapter 5. However, it is worthy of a reminder that a wide range of toxic plants exist in many parts of the world. If food is scarce, animals are more likely to consume less palatable plants, some of which could contain toxins. In extensive animal husbandry, water supplies are often unregulated and natural water courses or ponds allow livestock direct access to water of poor quality. This could be as a result of high mineral content or toxic cyanobacteria as a result of eutrophication (Silva et al., 2014). (Disease-causing organisms may also contaminate communal water resources through faecal contamination or sharing the resource with diseased livestock or wildlife.) A wide range of symptoms have been described following ingestion of toxins up to and including death (the latter usually due to a small number of large consumption events) but in some cases tolerance can be acquired. In addition to direct effects on livestock, secondary adverse effects in humans are sometimes anticipated.

Natural disasters such as earth-quakes, storm, or flooding can increase the animal disease risk in extensive systems in a number of ways. For example, livestock can injure them-selves in attempting to escape from local disasters such as wildfires or predator attack. When disasters strike, animals tend to be brought together (often as their keepers seek relief) increasing the risk of disease transmission, particularly around shared resources such as water troughs; if some form of biosecurity or quarantine can be established this is very valuable. Or animals can be exposed to novel pathogens if owners and their live-stock are displaced to new areas. Also, under disaster emergency conditions, animal medicines may not be available and large scale vaccination or disease control programmes can be disrupted. Under disaster management it is impor-tant to recognise that if many herders or other animal attendants and their stock congregate together, the risk of spread of zoonoses increases. For livestock, these zoonoses include anthrax, salmo-nellosis, tuberculosis, Rift Valley fever, rabies, and brucellosis. Rift Valley fever is spread by mosquitos and heavy rainfall can cause this vector to spread outside of its usual range, and thereby introduce disease to new areas. This spread can either be occasional or permanent if suit-able vector conditions persist. In some ways, livestock in extensive systems may be less susceptible to the effects of natural disasters as they are not usually heavily dependent on external inputs or infrastructure.

With increasing conflict around the world, along with the impact of climate change, livestock in extensive systems can also be exposed to man-made and natural disasters. Political conflicts have the capacity to displace large human populations and, where these people are engaged in livestock husbandry, maintenance of disease control and eradication strategies is very challenging and often not a top priority. Thrusfield (2005) lists 19 military campaigns that have disseminated rinderpest. More recently, in Southern Sudan, control of diseases including East Coast fever, foot-and-mouth disease, and trypanoso-miasis has been affected. Conflicts can affect primary animal production and put food security and livelihoods of pastoral communities at risk. In some areas of the world where there is much trans-boundary movement of livestock, war and civil unrest have the potential to disrupt wide-scale disease control and monitoring initiatives. As an historical example, the worldwide rinderpest eradi-cation campaign was seriously affected by the Second World War which resulted in resurgence of the disease in South and South-east Asia and, by the end of the 1940s, the disease was widespread in China.

A useful source of information in the event of an emergency or disaster involv-ing livestock is the Livestock Emergency Guidelines and Standards (LEGS, 2014). Here it is recognised that although humanitarian aid can be put into place quite quickly, it is important not to over-look the fate of animals on whose sur-vival and productivity millions of people depend worldwide. As well as cover-ing matters directly related to animal health applicable to extensive systems, other important aspects are covered such as ensuring that any restocking is done using animals of appropriate health status. The LEGS information also includes fascinating case studies of using both the traditional habbanaye method of providing keepers with animals who can keep the first offspring and then pass the original stock to the next beneficiary and livestock fairs where vouchers are used to buy animals. Training keepers in animal health can form part of these processes.

In extensive systems the main methods to control epidemic disease may be difficult or impossible to apply. Yet the impact of epizootic disease is often greatest amongst the poorest farmers and communities (Grace et al., 2012). Methods of control include enhanced biosecurity through:

- quarantine/movement restrictions for susceptible hosts;
- slaughter and safe disposal that does not, for example, encourage carcass removal by carrion;
- treatment/vaccination;
- disinfection;
- control of vectors/removal of exposure.

In extensive systems it can be particularly difficult to implement vaccination in good time in order for it to be effective. Yet vaccination programmes could lead to significant animal health impact in many areas where human–animal contact is infrequent. Even in non-epidemic situations, it is difficult to submit specimens to the laboratory and expect them to arrive in suitable condition in order to facilitate diagnosis.

Livestock in extensive systems are likely to interact with a range of wildlife species which may expose them to a number of diseases, especially in the case of the presence of wild ruminant species which co-exist with rangeland livestock; in many cases it is impossible to prevent contact (Figure 7.1). An example is given below but there are many others. In the case of vector-borne diseases, there is clearly no need for direct animal to animal contact.

A backdrop to all of the above is the challenge of technology transfer and increasing the skills of stockpersons working in isolation or in remote areas.

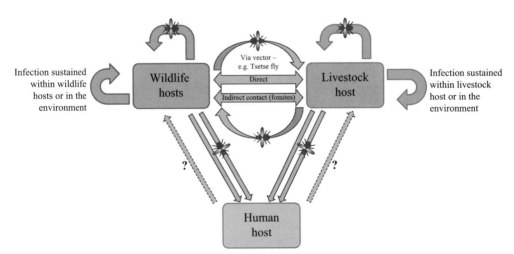

Figure 7.1: Stylized diagram showing the main interrelationships between wildlife, livestock, and human hosts for diseases affecting multiple species

There are many factors, which are often disease-specific, that can affect the "strength" of the arrows in the diagram or the likelihood of the routes existing at all or being important. Many of these are density-dependent factors which ultimately affects the "fitness" of the infectious disease organism, captured in the composite term R0. These include (but not in any order of importance):

- Density and structure (e.g., ages) of the host and recipient populations;
- Numbers of susceptible, infected, and recovered (SIR) animals in each population;
- Proportion of vectors which become competent in disease transmission;
- Control strategies in the (usually) livestock host (e.g., vaccination);
- Control strategies on vectors;
- Persistence of the pathogen in the environment;
- Variation in resistance between individual hosts;
- Pathogen genetic heterogeneity (ability to avoid host's specific immunity);

- Incubation periods in various hosts;
- Whether the vector is a simple "vehicle" for the organism or whether anything happens in the vector to alter pathogenicity or duration of infection – so called non-linear processes;
- Seasonal factors affecting vectors, particularly temperature and moisture;
- Mortality rate and time to death (diseases leading to rapid death with little time for onward transmission may fade out);
- Rate of replication of infectious agents in the host;
- Development of carrier state;
- Relative host susceptibility;
- Ability of disease to be maintained in wildlife or livestock hosts;
- Spatial distribution/functional overlap of wildlife and livestock and landscape features.

Less is known about the reinfection of either wildlife or livestock from humans, though treatment of human disease will reduce this risk.

Information needs to be presented in a form and at a time appropriate to the recipient if it is to have any durable effect (Goddard, 1999).

Mitigating Factors for Disease Risk

A number of generic factors can be considered to mitigate disease risk. Unfortunately, veterinary or extension services able to provide both immediate and long-term support may be patchy or non-existent. However, it is well recognised that stockperson knowledge as the basis of appropriate animal care is a key part of reducing both the risk of disease occurring and the impact if disease does appear. Recent technological advances provide a great opportunity for enhancement in this area and will be identified in the final section of the chapter.

Keeping appropriate breeds, particularly those adapted to the prevailing eco-climatic conditions, is clearly of fundamental importance and almost so obvious as not to warrant mention. It is, nevertheless, important to recognise that importing high-performing, immunologically naive individuals into situations where certain endemic diseases exist and nutritional resources will not support high levels of production are unlikely to be successful, however well meant. Encouraging stockpersons to use livestock adapted to local conditions is crucial, with the possible selection of breeds or individual lines with enhanced production characteristics, a process that could be speeded up using modern genomic techniques.

The net stocking density in many extensive systems is often thought of as low, given that even large numbers of animals have access to very large areas of grazing. On this basis, the transmission of disease between individuals due to proximity might be anticipated to be reduced or parasite loads diluted. Some herding systems move animals to different areas of grazing on a daily basis and this might also be thought to reduce the risk of ingesting infectious stages of certain parasites. However, it is important to remember that ruminants are often very social animals and will seek the close company of their conspecifics. For example, small ruminants seeking to mitigate a high ambient heat load will adopt a bunching strategy in order to achieve an overall net benefit (when sheep are unable to find shade they tend to crowd in order to reduce heat gain from radiation and increase heat loss by convection and conduction) or seek shelter from the sun under what shade is available.

Another example: to reduce the risk of night-time predator attack, animals from a rangeland may be corralled at night. While this also permits the inspection of livestock, the close proximity allows opportunity for disease spread. Thus even though a large area of grazing land may be available, animals will still associate closely and create "hot spots" for disease transmission to occur. Other examples include congregation around fixed feeding stations and watering points. (Indeed, turning off the water supply is sometimes used as a means to gather animals who will move to the watering point.) Certain disease vectors, ticks for example, have adapted to this and can be found preferentially where animals congregate.

Many extensive systems have thousands of head of livestock. Such a large single epidemiological unit can present almost insurmountable problems if an outbreak of infectious disease occurs. There may be no way to segregate infected from potentially uninfected animals and disease such as foot-and-mouth disease can rapidly sweep through all of the susceptible animals. This risk may be exacerbated by the stockperson's inability to inspect individual animals on a regular or consistent basis and so early cases of disease may be missed; any delay can have an escalating effect on the course of the disease outbreak.

For the nomadic stockperson, there is clearly a challenge in keeping abreast of local disease risk, enhancing skills, and obtaining any required medications. Thus the treatment of animals in some extensive systems can be very much a matter of traditional remedies of unproven efficacy. However, weaning traditional grazers off long-practised methods can only be ethically pursued if there is the possibility of long-term support for validated alternatives. Thus in many situations, stockpersons and herdsmen are naturally of a conservative nature and will prefer a cautious approach when they have found this reliable in the past.

EXAMPLES OF SOME SPECIFIC DISEASES

As noted earlier, it is not the intention to cover all diseases and other health risks that might afflict livestock in extensive systems but it may be helpful to provide some illustrative examples, where the extensive nature of the husbandry may have an impact, including where new developments have led to improved disease control, and in one case, total eradication.

EAST COAST FEVER

Also called *theileriosis*, it is caused by the intracellular protozoan pathogen *Theileria parva*. It is usually characterised by high fever, swelling of the lymph nodes, dyspnoea (difficulty breathing) due to pulmonary oedema, and high mortality. It is found characteristically in east, central, and southern Africa. It is transmitted during feeding on cattle by the brown ear tick, *Rhipicephalus appendiculatus*. Disease leads to severe economic consequences for pastoral cattle herders and small-scale keepers alike whose livelihood may depend on small numbers of individual animals. As there has been difficulty in developing

an effective vaccine and one that did not rely on an unbroken cold chain[2] to maintain efficacy, treatments generally relied on acaricides to control ticks. However, the need for frequent dipping, the difficulty of providing dips and spray races in remote areas, tick resistance, and potential toxicity for workers have led to acaricides being an unviable option in many cases (and the dips themselves can be environmental pollutants). Unfortunately ticks transmit diseases other than ECF so dipping or spraying with acaricides may still be required in these cases. Animals that recover from East Coast fever are immune to subsequent challenge with the same strains of *T. parva* but may be susceptible to some heterologous strains and most recovered (or immunized) animals remain carriers of the infection. Specific treatments have been advocated, for example using a combination of the antitheilerial compound parvaquone and the diuretic compound frusemide (Musoke et al., 2004) but their field application seems limited. In recent years, a vaccine has become available which involves infecting animals with a specific dose of live parasites and providing simultaneous treatment (the "infection-and-treatment method" ITM) – though a cold chain is still required until the last 6 hours (Di Giulio et al., 2009) and has become widely used, for example, in Tanzania. Around a million animals have been effectively vaccinated since 1998. Significantly, in relation to extensive systems, the majority of these

---

[2] A cold chain is essentially a delivery process whereby the integrity of many vaccines (usually) is only assured if the vaccine remains below a critical temperature (but usually not frozen) at all times between leaving the production facility and the point of use. This can be extremely difficult to achieve where distribution networks are long, have many steps, and where power supplies for refrigeration can be unreliable.

were from pastoral systems. As a result we can hope to see a significant reduction in the impact of this disease and therefore an improvement in livelihoods, providing adequate funding can be provided to support poorer farmers.

Sheep Scab

This is an acute or chronic dermatitis caused by a parasitic mite, *Psoroptes ovis* which causes intense, debilitating irritation due to an allergic, hypersensitivity reaction by sheep primarily to the mites' faeces and the general activity of the mites. This leads to loss of condition, secondary infections, and even death. The disease is highly infectious and mites can live off the host for up to 17 days. Not all sheep in a flock may show symptoms but it is likely that all sheep in a diseased flock are affected. In the UK, sheep scab was eradicated from the national flock in 1952 but was inadvertently re-introduced in 1973. Compulsory dipping of sheep started thereafter but was ended in 1992 and the burden of responsibility passed from government to farmers. Sheep scab is the most important ectoparasite disease of sheep in the UK. In Scotland, an industry-led Scottish Sheep Scab Initiative was introduced in 2003 which aims to reduce the incidence of sheep scab through flock biosecurity (part of which involves identifying risk) and co-ordinated treatment (an important strategy as noted earlier). The principles of this initiative are applicable to a wide range of diseases:

- Promoting and supporting best practice in flock biosecurity;
- Minimising the impact of outbreaks through effective and co-ordinated treatment;

- Maximising the effects of preventative action through targeting risk and co-ordinated treatment.

Where the availability of labour to aid in gathering sheep over wide areas (often of challenging terrain) is reduced (as is currently occurring in upland Scotland), it becomes increasingly difficult to gather all animals to apply treatments. Also, reduced presence of the stockperson means that biosecurity measures such as maintaining sheep-proof fences are not easily done. Failure to treat the whole flock can lead to resurgence of disease when the protection afforded by the treatment wears off. Persistent effort will be needed to achieve eradication again. This is likely to be aided by the application of a new diagnostic serological test which detects host antibodies to a specific protein, able to identify animals in the early stages of the disease before clinical signs become apparent. Research is underway to develop a vaccine which would have tremendous benefits to sheep production, especially in extensive flocks where it is easy for individuals to evade treatment and perpetuate the disease.

Animal Trypanosomiasis (Nagana)

This occurs in Africa and is spread by a number of tsetse fly (Glossina) species. *Trypanosoma congolense* and *Trypanosoma vivax* are the two most important species which affect cattle in sub-Saharan Africa. In cattle there is a reduction in the growth rate, loss of milk production, and general malaise which can lead to eventual death. The cause of over 95% of human African trypanosomiasis (HAT: sleeping sickness, a major human health concern) is *T. brucei gambiense* which can also affect cattle. *T.b. rhodesiense* is more commonly seen as having a domestic and wild animal reservoir but causes less human disease. Thus as domestic and wild animals are important in the epidemiology of the human disease there is added value in developing effective fly control methods (antigenic variation in trypanosomes makes vaccine development problematic). Tsetse flies occur in sub-Saharan Africa in "fly belts" where numerous fly control programmes have been deployed effectively over a number of years. As well as aggressive aerial spraying campaigns, there has also been more localised trapping of flies using chemical attractants as lures. Trap barriers have also been used to prevent reinfestation by tsetse flies (Schofield and Kabayo, 2008). The test of the use of laboratory-reared irradiated (and hence sterile) male tsetse flies of the *Glossina austeni* species (the sterile insect technique: SIT) was successful in causing their elimination on the island of Zanzibar (Vreysen et al., 2000) and this approach is now part of co-ordinated fly control activities that use all available techniques in mainland Africa. In some areas, an effective part of trypanosome control is through treating cattle with trypanocides and/or insecticides, based on improving understanding of tsetse behaviour (Hargrove et al., 2012). Thus livestock keepers can play a significant role in the control of human disease. Certain native species of cattle are *trypanotolerant* and can survive and grow even when infected with trypanosomes. Thus any stock which are moved to these areas should have this trait, rather than importing

European cattle with no resistance, or zebu breeds that are less resistant.

The Rinderpest Success Story

Rinderpest (cattle plague), a contagious viral disease of cattle and buffalo was once the scourge of extensive cattle production in Asia, Africa, and Europe where losses could reach 100% and hundreds of millions of cattle were killed by it. It also affected other cloven hoofed species but with a lesser pathogenicity; some wild species can carry the virus. It brought devastation to domestic and wild animal populations and where people's livelihoods depended on them, entire local or national economies. One of the significant areas of impetus for the establishment of the World Organisation for Animal Health (OIE: www.oie.int) in 1924 was an outbreak of rinderpest in imported animals in Belgium in 1920. It was perhaps fitting that the announcement of the eradication of rinderpest in 2011 coincided with the 250th anniversary of the establishment of the world's first veterinary school in Lyon, France, which was in fact created to help control rinderpest in Europe (Thrusfield, 2005).

The eradication of rinderpest, through international cooperation and concerted scientific and on-the-ground national veterinary services and animal owners' effort, has been one of the world's most significant disease control success stories yet this was only declared the case in May 2011 by the OIE. It is only the second infectious disease to have been eradicated (the other being smallpox). However, following eradication of the disease from livestock and wildlife, the OIE cautions against the continued storage of organisms (virus stocks, vaccines, and tissue samples which may contain viable virus) as susceptible animals are now vulnerable to infection through accidental or deliberate release. This eradication has had a hugely positive impact on cattle production in extensive systems in particular. Control was aided by the fact that following exposure and recovery (though uncommon) or vaccination with attenuated virus, lifelong immunity is established. Vaccination was deployed as part of a Global Rinderpest Eradication Programme (GREP), initiated in the 1990s, though immunisation of varying efficacy and safety was available from the early twentieth century. In 1960 the development through tissue culture of a rinderpest vaccine (TCRV) by Walter Plowright and co-workers that was highly effective, stable, safe, and cheap and later a vaccine that was not heat labile (thermostable), and thus easier to deploy in the field without extensive cold-chain maintenance, were key parts of the eradication effort. Thus again, producing vaccines which did not require a cold storage chain was a big element here and aided the ultimate elimination of a major disease.

Enhanced diagnostic tests were also crucial in the final struggle to defeat the disease. The last large outbreak occurred in Asia in 1994, brought by buffaloes imported into northern areas from the Punjab, resulting in huge losses in cattle, yaks, and buffaloes. (For more information on the eradication of rinderpest, see Roeder and Rich [2009] and Roeder et al., [2013].)

Brucellosis

Brucellosis is an infectious disease of ruminants that leads primarily to reproductive problems including abortion and is also a zoonosis which can have life-long consequences in infected humans. In many countries it has been eradicated through the implementation of testing and vaccination programmes. However, as with many diseases in extensive situations, the presence of a wildlife maintenance reservoir can make control and eradication in co-habiting livestock difficult or impossible. Apart from direct veterinary considerations, there may be important human dimensions to consider as exemplified by the current disease situation due to Brucella abortus in the Greater Yellowstone Area (GYA) in North America. Here, with an on-going brucellosis problem due to infection in elk (deer) and bison (buffalo), there are significant and long-term societal conflicts between individuals and agencies with different agendas and objectives (primarily protecting the health of livestock vs a sustainable elk and bison population). This constitutes a wicked policy problem; unless these tensions are recognised and properly managed through multi-stakeholder conversation there will be ongoing difficulties in disease management and the associated animal welfare cost.

Brucellosis was all but eliminated from cattle in the Greater Yellowstone Area (GYA) around the year 2000. This was as a result of an effective vaccination and serological monitoring programme. However, brucellosis is present in the bison and elk which inhabit the Yellowstone National Park and there is potential transmission of brucellosis from these wildlife reservoirs back to cattle, primarily by cattle being exposed to infected aborted material, though the abortion rate in bison is not high. The relative importance of elk or bison is not entirely clear and will relate to their spatial distribution around the time of birth (see Roberts et al., 2012). Elk tend to mix with cattle to a greater degree than do bison and are probably much more important in maintaining cattle infection (Rhyan, et al., 2013). In addition, management methods can reduce contact between bison and cattle. Another player in the ecosystem which may affect the spatial overlap between bison and elk is the wolf, following reintroduction in 1995. While within-species infection occurs in both elk and bison, bison (which have a high seroprevalence) are only thought to infect elk rarely (Proffitt et al., 2010).

Brucellosis was thought to have been introduced to bison by infected cattle early in the twentieth century (Dobson and Meagher, 1996). With measures implemented particularly around the 1970s, the population of bison has been increasing. The carrying capacity for bison in the National Park based on food resources has been estimated by Plumb et al. (2009). While the current population is below the estimate, factors such as harsh winters and pressure from competitive grazers such as elk encourage bison to seek winter grazing and spring calving grounds at lower altitudes outside the park (trans-boundary migration), places where livestock are ranched for part of the year, leading to the opportunity for mingling with livestock (if separation cannot be maintained) and disease transmission. Where elk and bison do move to these livestock areas, one option is to use helicopters and other vehicles to push them back to the park in springtime, a process known as hazing. This has provoked much public debate, not only in relation to the process itself but also, for

example, wider concern as some claim that the helicopters disturb the threatened grizzly bear population. Control of the disease in wildlife is key. Some bison are herded up, tested for brucellosis and test positive animals killed, and there is also an encouragement of hunting bison near park boundaries. Practices which cause elk to congregate, such as winter feeding should be avoided. Vaccination of wildlife against brucellosis using remote delivery is also being considered as an option. Generally, though, vaccination alone is unlikely to eliminate any infectious disease.

Given the wide range of interests and the often conflicting views, an Interagency Bison Management Plan (IBMP) was implemented in 2000 following negotiations between the National Park Service and others, including the Montana Department of Livestock, and was based on principles of adaptive management. Adaptive management is the process by which plans are not fixed but can evolve based on an improving evidence base and continuing stakeholder input. Round-table groups that consider the disease transmission risks and related population dynamics and conservation status of both elk and bison represent a good way to encourage all views to be aired and the public to feel they have a stake in the decision-making process.

Bovine Tuberculosis

As a final example, consider bovine tuberculosis. Bovine tuberculosis (bTB) is a significant zoonosis occurring around the world and there is a considerable effort to reduce its incidence. In relation to the importance of livestock–wildlife interactions, it is viewed as a (re-)emerging disease of European wildlife and hence a risk for domestic livestock grazing extensive areas where infectious organisms can be found. A recent report (Fink et al., 2015) has shown that red deer can act as a maintenance host for bovine tuberculosis in some alpine regions of Europe. The specific organism under investigation was *Mycobactrium bovis* ssp. *caprae* and the authors concluded that in certain hotspot areas where large numbers of deer congregate (for example around feeding sites), they could act as a source of infection for domestic cattle when grazing the following summer (as well as further infection of conspecific deer), rather than being simply spillover hosts.[3] (Molecular studies showed previously that the isolates from deer and cattle were the same genotype; Prodinger et al., 2002.) This latest study advances previous observations that in a different part of Europe, *M. bovis* and *M. caprae* were not detected but suggested that increasing population numbers of wild ungulates and management that causes wild species to congregate may represent a risk (Schöning et al., 2013). An example from the USA shows how a proactive wildlife control programme can effectively prevent the establishment of a wildlife disease reservoir (Carstensen and DonCarlos, 2011). Note that other principal hosts of bTB and implicated in livestock disease are the badger in the UK and the brush-tailed possum in New Zealand. Interestingly, terminally diseased badgers and possums have atypical behaviour (such as loss of fear towards cattle and deer) which may enhance disease spread (Muirhead et al., 1974; Sauter and Morris, 1995).

CONTROL OF HEALTH AND DISEASE IN EXTENSIVE SYSTEMS

In the previous section, examples of some specific diseases which have the ability to be particularly problematic in extensive systems were described. The intention now is to note some generic issues for the control of health and disease. For example, many diseases are vector-borne and the target for disease control is often the vector. Tick control through the use of acaricides (though now significantly compromised by the emergence of resistance in ticks), the evolution of new ectoparasiticides and a range of vaccines have been of major importance to animal health and consequent human hunger and poverty alleviation. However, this is a biological arms race as resistance to acaricides in Africa in particular has shown. Understanding of the life cycle of vectors and the key points of interaction with disease hosts is constantly improving.

Gastrointestinal parasites are a significant cause of disease worldwide, with an accompanying loss of animal performance. The use of anthelmintics to control parasitic nematode worms has become widespread, including in many extensive livestock systems and, historically, this has been hugely successful. Now the emergence of resistance, even against some novel compounds, may begin to

---

[3] A spillover host is a host which becomes infected but generally plays no further role in onward transmission. However, on occasions, diseases can "spillback" to the original host (see Nugent, 2011).

limit the ability to keep livestock in some situations. It is crucial that farmers monitor the resistance status of nematode populations and hence preserve the efficacy of the compounds they are using and apply best management practice to limit the further development of worm resistance. In the UK, the Sustainable Control of Parasites in Sheep (SCOPS: http://www.scops.org.uk/) initiative provides a wide range of up-to-date advice on best practice that can be adapted for use in many production systems.

Under field conditions, vaccines which rely on cold storage to maintain efficacy are difficult to deliver in a reliable way. So in parallel with the use of vaccines there is considerable effort expended to seek control measures which are likely to be durable and easy to apply. Many of these can focus on the control of insect vectors through environmental measures such as improved water management in the broadest sense to remove breeding grounds. This can have the added benefit of reducing the need for chemical treatment with the accompanying environmental benefits.

The importance of the presence of wildlife reservoirs of livestock disease in terms of maintaining disease and facilitating spread has already been noted (Figure 7.1). Rhyan and Spraker (2010) identified that many problems in relation to diseases emerging from wildlife occur when humans encroach into wildlife habitats. Pressure on land use will shift this dynamic further in the future. As pointed out by Miller et al. (2013), the complex nature of livestock–wildlife interactions requires a thorough understanding of the relevant disease epidemiology, transmission, and maintenance in order for disease processes to be properly understood. In addition, successful control of such diseases generally requires a multidisciplinary approach (such as One Health).

Application of biosecurity and biocontainment are a particular challenge in extensive systems. Coupled with this, the application of flock/herd planning is perceived to be poorer in extensive systems (where we might find fewer factors under control) but it is arguably more important. In many countries around the world, the availability of veterinary services to farms or migratory animals in remote areas is problematic, especially given the relatively poor financial returns of many such enterprises, where support through subsidies is the only way they can continue to exist. The Emergency Centre for Transboundary Animal Diseases (ECTAD) at the Food and Agriculture Organization of the United Nations (FAO) has responsibility and the capacity for detection and prevention of threats caused by zoonotic and high impact diseases on a global scale. In the face of disease outbreaks, the availability of suitable antibiotics can limit their impact and extent.

There are now many national and international programmes which aim to tackle some major livestock and zoonotic diseases but the provision of local veterinary and para-professional resources dealing with animals on the ground remains crucial to improving the general level of extensive livestock health.

ONE HEALTH AND THE RISK OF ZOONOSES

The One Health paradigm embraces the consideration of interdependence of human, animal, and environmental health.

A review of the progress made through this approach and a synthesis of achievements over the first ten years after the concept was developed has been given by Gibbs (2014). There are numerous historical examples of the close relation of disease in livestock and humans but we are now becoming better aware of the mechanisms by which diseases can overcome inter-specific barriers, including the role of genetic mutations. The rapidly changing landscape of extensive livestock–wildlife interactions in many countries can shift the balance in hosts which carry zoonotic diseases and requires new strategies to minimise the occurrence of these diseases in humans. In addition there seems an increased risk of disease agents jumping species from wildlife to livestock and humans.

Control of zoonotic disease can be a particular problem in developing countries where there is a high level of extensive livestock production. It has been estimated that 75% of emerging infections[4] are zoonotic (Taylor et al., 2001) so there is a clear motivation to direct particular focus in this area worldwide. Endemic zoonotic diseases include brucellosis, Q-fever, (bovine) tuberculosis, anthrax, and Rift Valley fever, but there are many others. There is likely to be under-reporting or failure to diagnose these, especially in the poorer extensive livestock areas. Some of the symptoms of such zoonoses make definitive diagnosis difficult but they have a major toll on many people and communities. For example, *T. brucei rhodesiense* is a zoonotic cause of human African sleeping sickness which, as for a number

of other diseases, may not necessarily cause symptoms in the animal host. Keeping animals healthy is a crucial way to reduce the likelihood of zoonotic disease in extensive systems.

Increasing recognition is being focussed through the One Health initiative which shines a light on the parallel between human and animal disease and aims to integrate aetiological and epidemiological approaches in livestock and humans in a holistic way (for a recent review, see Halliday et al., 2015). The One Health principles also involve improved monitoring as part of disease risk mitigation. (In the case of zoonotic disease this may also involve modifying the way humans and animals interact.) A key feature is through enhanced collaboration at all levels. Mobile phone technology could allow the identification of zoonotic disease hotspots. The One Health approach has recently been advocated as being key to the control of tuberculosis in terms of protecting humans, livestock, wildlife, and the environment (Miller and Olea-Popelka, 2013).

LOOKING AHEAD: WHAT ELSE DO WE NEED TO KNOW ABOUT THE HEALTH OF LIVESTOCK IN EXTENSIVE SYSTEMS?

A number of worldwide issues are likely to impact on the health of extensively-managed livestock in the coming decades, harbingers of which are already seen. For many traditional upland farmers there is often dependence on single

---

[4] Emerging infections are defined as those resulting in new or unusual syndromes or infections or increases in epidemic disease.

individuals looking after large numbers of animals in remote areas. These often older farmers are not immune from societal change; succession by the younger generation is a major problem and the often lonely and hard existence (in terms of weather and terrain) is seen as unattractive. Importantly in this context it becomes increasingly difficult to be aware of and deliver modern animal health programmes. These challenges can lead to a downward spiral amongst a group of farmers who might share a common grazing resource and who cooperate to deliver periodic health care and disease monitoring: as one drops out, life becomes increasingly difficult for the remainder. This has wider implications for local communities as the recent report on farming's retreat from the hills (SAC, 2008) shows in relation to sheep and cattle numbers in a Scottish hill and upland context. (Even one of the most revolutionary changes for sheep farmers, the arrival of the four-wheel-drive motorbike, will be unlikely to mitigate the impact of this demographic change). Such a change can be seen also in some traditional Inuit and Sami reindeer hunting and herding communities, where declining economies of traditional husbandry and the need to produce meat of a high health status also come into play.

On the opposite hand, precision livestock farming (PLF) is being increasingly adopted. PLF defines the use of optimal technology (for example close modelling of input–output relationships) to make best use of all available resources often through targeted improvements and the reduction of waste. While this may be thought of as an approach not suited to extensive systems, there are lessons to be learnt. For example, there could be benefits from optimising the management of health of animals through being more precise about targeting prophylactic measures. Early warning of parasitism through failure to meet weight targets is a case in point. While capturing information remotely from animals using digital technology is generally seen as something for non-extensive systems, perhaps, in due course, this will be a potent resource for extensive systems to allow the stockperson to keep in touch with remote animals and monitor important aspects of health (and welfare). For example, motion sensing could identify lame animals or those with injuries preventing movement which need to be found and treated.

Climate warming has the potential to influence the occurrence of a number of diseases, often through altering (usually increasing) the geographic presence of vectors. This may have contributed to the emergence of Bluetongue in northern Europe. The midge, *Culicoides imicola*, is the important vector of Bluetongue virus in Africa and southern Europe but local culicoides species served as the vector of Bluetongue serotype 8 which emerged in northern Europe in 2006 (the route by which Bluetongue arrived has never been definitively identified). The worldwide movement of humans, animals, and disease organisms has great propensity to bring disease to new areas and livestock populations and it is essential that information about risks can penetrate all livestock-keeping areas. The ability to provide information, extension advice, and practical support to stockpersons who care for livestock in remote areas is a challenge that needs to be better addressed as the transfer and uptake of new technology and ideas is undoubtedly more difficult given the geographic dispersal of extensive systems.

The development of animal-side tests could allow testing and instant treatment/removal for animals that can be gathered only occasionally. One of the first such tests was developed by the Pirbright Institute in the UK as part of the rinderpest control programme and used eye swabs, mixed with reagent, applied to a test slide containing antibody-coated latex beads to test for the presence of rinderpest virus based on an immune-capture ELISA. There are now pen-side tests for other pathogens such as foot-and-mouth disease virus. Molecular diagnostic approaches will also improve diagnostic resolution and be better able to fingerprint disease outbreaks and enhance epidemiological studies. Such knowledge should also enhance the targeting of disease control programmes, though with so many potential diseases, particularly in developing countries, resources cannot currently support sufficient activity. Part of the enhanced knowledge availability relates to identification and traceability, which digital technology allows.

Digital or "smart" technology is currently seen in action in a range of non-extensive systems. Some of the first direct animal applications were in individual dairy cow identification through devices in collars to allow animals to receive individual rations in the milking parlour. Now electronic identification (EID) is used in many settings, not just to provide official unique identifiers. Automated weighing and handling systems are available for sheep. Movement sensors and associated computer algorithms can be used for heat (oestrus) detection and potentially identifying lame animals. For example, it is now possible to send pictures and a description from a smartphone to a regional diagnostic centre. Along with images, the sender automatically provides accurate GPS information about their location so emerging epidemics can easily be graphically-linked and hence mapped. We are also soon likely to see "mini labs" into which samples are placed and which interface results via the smartphone to a diagnostic centre. This takes advantage of biomarker discovery ("pathogen-specific fingerprints") which require minimal sample processing steps through the identification of disease-specific molecules. Such developments in mobile phone-connected diagnostic tests have a huge potential to improve disease recognition and control for livestock in resource-limited, extensive systems. In particular, providing early warning of epidemic disease allows rapid and targeted intervention. This will also enhance the collection of "big data" for global animal health. Much information is collected and available through the internet disease reporting system, ProMED (Programme for Monitoring Emerging Diseases) which primarily focusses on human disease or diseases in humans where animals may be involved. In the opposite direction of information flow, disseminating information about local disease threats or extreme weather prediction via rural animal health workers is becoming increasingly valuable. This will facilitate the early identification of disease and more integrated animal identification databases should enhance traceability in the case of a disease outbreak.

Animal health in extensive systems should also be considered in the light of any potential impacts on food security and food safety. The availability of veterinary resources to support this need to ensure optimum health status is sometimes poorer in areas where extensive systems are found. Every effort should be made to use modern technology to deliver enhanced standards of care. The further provision of para-professional animal health workers ("barefoot" vets, para-vets, vet scouts, community animal health workers) is one way to move in this direction. Having barefoot workers in each community, trained in a range of basic veterinary skills, first aid, and vaccination also gives scope for them to be involved with wider-scale disease control programmes and be the conduit for up-to-date information between government veterinarians and the community. Even for nomadic systems where animals and their stockpersons are remote from any state veterinary services, this is an effective way to improve primary healthcare. Studies undertaken of this provision have shown its effect to be positive on the wellbeing and economies of poor farmers (e.g., Martin, 2001).

From the above it can be seen that there are many recent technological advances which have the tremendous potential to have a positive impact on the health of livestock in extensive systems. Over a relatively short period it is likely that many of the once remote and unengaged pastoralists can be brought into the mainstream of knowledge exchange and be better able to control important diseases in their livestock using the most effective technological methods. Such approaches will also allow engagement with extensive livestock keepers around the world in order to maximise the health of their animals, thereby increasing production, reducing their

climate change impact and reducing the likelihood of zoonotic disease in humans. Coupled with existing best practice, all of these innovations, the use of strategic vaccination, and improved infrastructure (e.g., better water storage and irrigation, and waste management so as to reduce the breeding potential of arthropod vectors) need adequate resourcing. The fact that the most important diseases of extensive livestock, many of which are zoonotic, are generally found in poorer areas of the world means that considerable and sustained external funding will be needed to realise the potential for disease control.

REFERENCES

AVMA (2008) https://www.avma.org/KB/Resources/Reports/Documents/one-health_final.pdf

Black, H., Alley, M.R., and Goodwin-Ray, K.A. (2005) Heat stress as a manageable risk factor to mitigate pneumonia in lambs. *New Zealand Veterinary Journal* 53, 91–92.

Bock, R.E., de Vos, A.J., Kingston, T.G., and McLellan, D.J. (1997) Effect of breed of cattle on innate resistance to infection with *Babesia bovis, B. bigemina* and *Anaplasma marginale. Australian Veterinary Journal* 75, 337–340.

Carstensen, M., and DonCarlos, M.D. (2011) Preventing the establishment of a wildlife disease reservoir: A case study of bovine tuberculosis in wild deer in Minnesota, USA. *Veterinary Medicine International* Article ID 413240, 10 pages, doi:10.4061/2011/413240

Collins, K.H., and Weiner, H.S. (1968) Endocrinological aspects of exposure to high environmental temperatures. *Physiological Review* 48, 785–794.

Di Giulio, G., Lynen, G., Morzaria, S., Oura, C., and Bishop, R. (2009) Live immunization against East Coast fever—current status. *Trends in Parasitology*, 25, 85–92. doi: 10.1016/j.pt.2008.11.007

Dobson, A., and Meagher, M. (1996) The population dynamics of brucellosis in the Yellowstone National Park. *Ecology* 77, 1026–1036.

Fink, M., Schleicher, C., Gonano, M., Prodinger, W.M., Pacciarini, M., Glawischnig, W. et al. (2015) Red deer as maintenance host for bovine tuberculosis, Alpine Region. *Emerging Infectious Diseases* 21:3 http://dx.doi.org/10.3201/eid2103.141119

Gibbs, E.P.J. (2014) The evaluation of One Health: A decade of progress and challenges for the future. *Veterinary Record* 174, 85–91.

Goddard, P.J. (1999) Information and technology transfer in farm animal welfare. British Society of Animal Science, Occasional publication 23, 75–82.

Grace, D., Mutua, F., and Ochungo, P. (2012) Mapping poverty and likely zoonoses hotspots. *Difid Zoonoses Report* 4, 1–119.

Gregory, N.G. (2007) *Animal welfare and meat production.* CABI, Wallingford, UK.

Halliday, J.E.B., Alan, K.E., Ekwem, D., Cleveland, S., Kazwala, R.R., and Crump, J.A. (2015) Endemic zoonoses in the tropics: A public health problem hiding in plain sight. *Veterinary Record* 176, 220–225. doi10.1136/vr.h798

Hargrove, J.W., Ouifki, R., Kajunguri, D., Vale, G.A., and Torr, S.J. (2012) Modeling the control of trypanosomiasis using trypanocides or insecticide-treated livestock. *PLoS Neglected Tropical Diseases* 6, e1615. doi:10.1371/journal.pntd.0001615

Hodgson, J. (1990) *Grazing management: Science into practice.* Longman, Essex. 203 pp.

LEGS (2014) *Livestock Emergency Guidelines and Standards*, 2nd edn. Practical Action Publishing Ltd., Rugby, UK. 314pp http://www.livestock-emergency.net/resources/download-legs/

Martin, M. (2001) *The impact of community animal health services on farmers in low-income countries: A literature review.* Vetaid, UK

Miller, R.S., Farnsworth, M.L., and Malmberg, J.L. (2013) Diseases at the livestock–wildlife interface: Status, challenges, and opportunities in the United States. *Preventive Veterinary Medicine* 110, 119–132.

Miller, M., and Olea-Popelka, F. (2013) One Health in the shrinking world: Experiences with tuberculosis at the human–livestock–wildlife interface. *Comparative Immunology, Microbiology and Infectious Diseases* 36, 263–268.

Muirhead, R.H., Gallagher, J., and Burn, K.J. (1974) Tuberculosis in wild badgers in Gloucestershire: Epidemiology. *Veterinary Record* 95, 552–555. doi:10.1136/vr.95.24.552

Musoke, R.A., Tweyongyere, R., Bizimenyera, E., Waiswa, C., Mugisha, A., Biryomumaisho, S., McHardy, N. (2004) Treatment of East Coast fever of cattle with a combination of parvaquone and frusemide. *Tropical Animal Health and Production* 36, 233–45.

Nugent, G. (2011) Maintenance, spillover and spillback transmission of bovine tuberculosis in multi-host wildlife complexes: A New Zealand case study. *Veterinary Microbiolog*, 151, 34–42. doi: 10.1016/j.vetmic.2011.02.023

Plumb, G.E., White, P.J., Coughenour, M.B., and Wallen, R.L. (2009) Carrying capacity, migration, and dispersal in Yellowstone bison. *Biological Conservation* 142, 2377–2387.

Prodinger, W.M., Eigentler, A., Allerberger, F., Schönbauer, M., and Glawischnig, W. (2002) Infection of red deer, cattle, and humans with *Mycobacterium bovis* subsp. *caprae* in Western Austria. *Journal of Clinical Microbiology* 40, 2270–2272.

Proffitt, K.M., White, P.J., and Garrott, R.A. (2010) Spatio-temporal overlap between Yellowstone bison and elk – implications of wolf restoration and other factors for brucellosis transmission risk. *Journal of Applied Ecology* 47, 281–289.

Rhyan, J.C., Spraker, T.R. (2010) Emergence of diseases from wildlife reservoirs. *Veterinary Pathology* 47, 34–39. doi: 10.1177/0300985809354466

Rhyan, J.C., Nol, P., Quance, C., Gertonson, A., Belfrage, J., Harris, L., Straka, K., Robbe-Austerman, S. (2013) Transmission of brucellosis from elk to cattle and bison, Greater Yellowstone Area, USA, 2002–2012. *Emerging Infectious Diseases*, 19 http://dx.doi.org/10.3201/eid1912.130167 [accessed 12.05.2015].

Roberts, T.W., Peck, D.D., and Ritten, J.P. (2012) Cattle producers' economic incentives for preventing bovine brucellosis under uncertainty. *Preventive Veterinary Medicine* 107, 187–203.

Roeder, P., and Rich, K. (2009) Conquering the cattle plague: The global effort to eradicate rinderpest. In: Spielman, D.J. and Pandya-Lorch, R. (Eds.) *Millions fed: Proven successes in agricultural development.* International Food Policy Research Institute (IFPRI). pp. 109–116. http://www.ifpri.org/sites/default/files/publications/oc64ch16.pdf

Roeder, P., Mariner, J., and Koch, R. (2013) Rinderpest: The veterinary perspective on eradication. *Philosophical Transactions of the Royal Society B* 368, 1623 doi: 10.1098/rstb.2012.0139

Sauter, C.M., and Morris, R.S. (1995) Dominance hierarchies in cattle and red deer (*Cervus elaphus*): Their possible relationship to transmission of bovine tuberculosis. *New Zealand Veterinary Journal* 43, 301–305.

Schofield, C.J., and Kabayo, J.P. (2008) Trypanosomiasis vector control in Africa and Latin America. *Parasites & Vectors* 1, 1–24. doi:10.1186/1756-3305-1-24

Schöning, J.M., Cerny, N., Prohaska, S., Wittenbrink, M.M., Smith, N.H., et al. (2013) Surveillance of bovine tuberculosis and risk estimation of a future reservoir formation in wildlife in Switzerland and Liechtenstein. *PLoS ONE* 8(1), e54253. doi:10.1371/journal.pone.0054253

Scottish Agricultural College (2008) *Farming's retreat from the hills.* SAC Rural Policy Centre.

Silanikove, N. (1992) Effects of water scarcity and hot environments on appetite and digestion in ruminants: A review. *Livestock Production Science* 30, 175–194.

Silanikove, N. (2000) Effects of heat stress on the welfare of extensively managed domestic ruminants. *Livestock Production Science* 67, 1–18.

Silva, A.C., Souza, A.M., and Dutra, I.S. (2014) Occurrence of blue-green algae in the drinking water of extensively raised cattle. *Pesquisa Veterinaria Brasileira* 34, Doi/10.1590/S0100-736X2014000500005

Squires, V.R., and Sidahmed, A. (1997) Livestock management in dryland pastoral systems: Prospects and problems. *Annals of Arid Zone* 36, 79–96.

Taylor, L.H., Latham, S.M., and Woolhouse, M.E.J. (2001) Risk factors for human disease emergence. *Philosophical Transactions of the Royal Society of London* 356, 983–989.

Thrusfield, M. (2005) *Veterinary epidemiology*, 3rd edn. Blackwell Science, Oxford.

Vreysen, M.J., Saleh, K.M., Ali, M.Y., Abdulla, A.M., Zhu, Z.R., Juma, K.G., Dyck, V.A., Msangi, A.R., Mkonyi, P.A., and Feldmann, H.U. (2000) *Glossina austeni* (Diptera: Glossinidae) eradicated on the island of Unguja, Zanzibar, using the sterile insect technique. *Journal of Economic Entomology* 93, 123–135.

WHO, 2014 (http://apps.who.int/iris/bitstream/10665/112642/1/9789241564748_eng.pdf?ua=1)

FURTHER READING

FAO, 2013. *World Livestock 2013–Changing disease landscapes*. Rome.
For examples of preparedness for animal disease emergencies, look at AUSVETPLAN: http://www.animalhealth australia.com.au/programs/emergency-animal-disease-preparedness/ausvetplan/

chapter eight

Neonatal mortality of farm livestock in extensive management systems

Cathy M. Dwyer and Emma M. Baxter

INTRODUCTION

For mammals, the most vulnerable period of an animal's life is the day of birth, with nearly 50% of pre-weaning mortality in cattle and sheep, and up to 20% in pigs, occurring within the first 3 days of life (Patterson et al., 1987; Nowak et al., 2000; Edwards and Baxter, 2015). At birth the foetus transitions from an environment where it is kept warm, protected from pathogens and environmental challenges, and provided with continuous nutrition and oxygen via the umbilical cord, to a relatively hostile extra-uterine environment. This is associated with a rapid adjustment of many physiological processes to enable breathing, to develop motor functions, maintain body temperature, and seek a food source (usually the maternal udder followed by sucking). The transition itself can also be a risk when the birth process is prolonged or difficult so constituting a threat to the survival of the neonate, with physical injury and hypoxia increasing with the duration that the animal is in the birth canal. Most newborn farm animals are born immunologically naive, so need to ingest colostrum (or "first milk," which is rich in immunoglobulins) to provide them with passive protection against environmental pathogens. Although maternal behaviour can facilitate these transitions, failure to accomplish this transition successfully is a significant cause of offspring mortality. In addition to the individual biological challenges faced by the newborn farm animal, it may be exposed to a number of external or environmental challenges (e.g., threats from predators, extremes of temperature, risks of rejection or crushing from its own mother, competition or aggression from other animals), that threaten its survival. Given these many challenges experienced within a few moments of birth it is perhaps not surprising that mortality of newborn livestock is a significant challenge for producers, and for the welfare of farm animals.

In indoor or intensive management systems, husbandry and attention from stockworkers can supplement maternal behaviour in ensuring a successful transition to extra-uterine life. The indoor

Animal and Veterinary Sciences, SRUC, Edinburgh, UK

environment can be closely controlled with respect to temperature and exposure to dangers, and additional nutrition or veterinary care can readily be provided. This may, however, come at the expense of some behavioural freedoms which may limit the expression of maternal care or neonatal behaviours, for example, parturient mothers might be confined in restrictive pens or crates (such as farrowing crates for sows which can have a negative impact on maternal behaviour; Jarvis et al., 1998), which reduce the opportunities for mothers to interact with their newborn offspring. Close confinement also requires attention to hygiene as a build-up of pathogenic load can occur where many mothers may give birth in close proximity. Thus in intensive management systems there may be an increased protection of offspring from thermal challenge, or threat of predation, but this is balanced against the increased risk of infectious disease and potential behavioural restriction. In more extensively managed systems the behavioural abilities of mother and young become more important in ensuring the survival of the young: although the risk of infection is reduced, there is a greater risk of other harms befalling the newborn animal, with less opportunity for human intervention to support survival. The welfare costs to the neonatal animal, and its mother, can then be affected by the degree of adaptation of the species or breed to their environment, and the degree of human contact or supervision that may be possible or available in different extensive environments. In systems with close shepherding, for example, animals may be afforded behavioural freedoms, but still benefit from human interventions in times of obstetric difficulty. In other systems, such as very extensive environments where the animals may be managed almost as wild animals, almost no human contact may occur at birth. For many births this will be of no consequence, and may even be beneficial, however, should there be complications then the mother and offspring are at risk of suffering catastrophic welfare costs. As discussed elsewhere, the role that the stockperson might play in the welfare of the neonate will be dependent on their skills, abilities, and experience in managing obstetric care and husbandry of newborn animals.

In this chapter, the potential risks for neonatal mortality in extensively managed livestock species, the potential welfare issues, and opportunities to improve survival will be considered. The main species that often give birth in an extensive, or outdoor, environment, and are the subject of this chapter, are sheep, beef cattle, and, in some countries, pigs. Where information is available on neonatal mortality in extensively managed goats, deer, and camelids, this is also considered.

NEONATAL MORTALITY AND ANIMAL WELFARE

Although mortality can be considered an unambiguous indicator of poor welfare, in that a system with high levels of neonatal losses obviously does not provide good welfare, there has been some debate about the actual welfare costs to the moribund neonate from the point of view of its subjective experiences (Mellor and Gregory, 2003; Mellor and Diesch, 2006). These authors postulate that, if mortality occurs during or immediately after birth, then the protective mechanisms of foetal life, which keep the foetus largely in an unconscious state, might also prevent the neonate from experiencing negative emotional states before death. If, however, the neonate establishes independent breathing, with the associated increase in oxygen saturation in the blood, then it is likely to be able to experience the same range of emotional states as older animals. However, many neonatal deaths do not occur immediately at birth but after a potentially prolonged period of attempting to establish independent feeding and thermoregulation, and even animals considered to be stillborn may have breathed prior to death (Barrier et al., 2013a), thus the welfare of the neonate is still an important issue.

The main subjective experiences of the moribund neonatal farm animal are considered to be breathlessness, hypothermia, hunger, sickness, and pain (Mellor and Stafford, 2004). Of these, from analogy with human experience and farm animal studies, hunger, sickness, and pain may be the most severe welfare challenges experienced by the neonate, as breathlessness and hypothermia can lead to a gradual loss of perception and awareness. Also of potential concern, although this has never been investigated, is the possibility that the maternal animal may experience frustration, anxiety, loss, and/or pain from a full udder, associated with the inability to show the full development of maternal care.

CAUSES AND RISKS OF MORTALITY IN DIFFERENT SYSTEMS

The main causes of neonatal mortality in different species, and the risks of experiencing the main welfare consequences of mortality, are influenced by species biology, and the common management practices associated with the species. With the exception of the pig, most farmed mammalian species produce between one and three offspring, and maternal care is focused on the individual. These offspring are largely protected from the need to compete with littermates for access to maternal resources (warmth and nutrition), but may need to show greater independence in establishing and maintaining contact with the mother than animals born into larger litters. Unique amongst farmed mammals, the pig builds a nest before giving birth, when given the opportunity to express these behaviours, which affords the newborn piglets some protection from thermal challenges and exposure to predators, and contributes to the quality of maternal behaviour expressed (Algers and Uvnäs-Moberg, 2007). Despite this protective environment, the large litter means each piglet has to establish itself at a teat in competition with others, and the relatively small size of piglets in comparison to the sow means there is increased risk of being trodden on or physically injured through crushing. The specific issues relating to each species will be considered in detail below, however, the main causes and welfare consequences of neonatal mortality common to all farmed species will be discussed here.

Hunger is one of the most important welfare challenges experienced by the neonatal animal, particularly those managed extensively. An inability to feed adequately after birth may be related to poor behavioural expression in the mother, for example unwillingness to allow the neonate to have access to the udder, or due to failure of the offspring to express appropriate activity or udder-seeking behaviour. In addition, the mother may be unable to produce a sufficient quantity of milk to feed all her offspring, and competition from littermates may induce missed sucking opportunities in those littermates that are less able to compete effectively for access. Although a complete failure to suckle might rapidly lead to hypothermia, apathy, and loss of consciousness, in many cases young farm animals are able to access some milk, but may be unable to gain sufficient nutrition to support good growth and development. Mortality, perhaps from other causes, might thus occur after a protracted period of hunger, weakness, and undernutrition. Although impaired sucking ability, inadequate lactation, or maternal interactions can prevent the offspring from ingesting sufficient colostrum to meet its nutritional needs, milk and colostrum are also often the only sources of fluid intake for the neonate. In high ambient temperatures, impaired milk ingestion can lead to dehydration and contribute to increased mortality from hyperthermia (Haughey, 1980; Stephenson et al., 1984).

In extensive management systems neonates are particularly vulnerable to thermal stress, with the greatest risk coming from cold stress or hypothermia. In sheep, for example, nearly half of all perinatal lamb losses are attributed to hypothermia, with cold, wet, or windy weather at lambing time causing a dramatic increase in mortality (Dwyer, 2008a). While newborn animals are wet with amniotic fluids, immediately after birth, their ability to withstand cold temperatures is severely limited and even

once dry their main method of producing heat is through brown fat metabolism, as they do not have sufficient muscle mass to generate much heat through shivering. In some species, particularly the pig, reserves of brown fat may be absent or rapidly depleted after birth, thus cold temperatures can result in the rapid onset of hypothermia and death. The additional vulnerability of the neonatal piglet is often addressed in their management. In contrast to their ruminant counterparts, extensively managed pigs are given straw-bedded farrowing arcs or huts, usually set in individual paddocks, in which to give birth and rear their young. This affords some thermal protection, isolation from the rest of the herd and a certain amount of protection from predators.

Neonatal animals may experience pain, injury, and death as a result of birth difficulty, which may be exacerbated in an extensive system where obstetric assistance from stockworkers may be delayed or impossible. Birth-related injury has been reported to play a contributory role in the deaths of 80% of neonatal lambs (Haughey, 1993), and is the major factor in beef calf mortality (Barrier et al., 2013b). Neonates can suffer a range of injuries, particularly involving haemorrhage around the brain and spinal cord, subcutaneous oedema, fractures, or rupture of the liver. By extrapolating findings from studies of central nervous system haemorrhages carried out in humans (Moussouttas et al., 2006; Schwedt et al., 2006), neonates experiencing bleeding around and into the brain as a result of birth difficulty are likely to experience severe pain. The relative size of the neonate in comparison to maternal size is a risk factor in birth-related injuries and for this reason piglets are rarely at risk. However, piglet size can influence farrowing duration and therefore stillbirth rate. In addition to the physical injuries experienced with difficult births, protracted labour increases the risk that the foetus will be without oxygen for a period of time resulting in hypoxia. In all species there is an optimal birth weight, with very heavy and very light neonates at greatest risk of stillbirth through hypoxia (Canario et al., 2006), although often for different reasons. Severe hypoxia can result in stillbirth, but additionally birth-injured and hypoxic neonates surviving the birth process will have low vigour (Dwyer, 2003; Baxter et al., 2008; Barrier et al., 2012), and an impaired ability to produce heat (Bellows and Lammoglia, 2000). These neonates can therefore also be vulnerable to other welfare challenges and have reduced postnatal survival.

Other sources of offspring injury can arise through their interactions with their mothers, or other animals in the social group. Failure of mothers to show an adequate expression of maternal care can result in aggression towards their own neonates, particularly in females giving birth for the first time. In extensive management systems, where animals have more behavioural freedom to express species-typical preparturient and birth behaviours, this may be less frequent than in intensive systems where animals experience greater restriction or social stress. Certainly in pigs, there is evidence that the behavioural restriction placed on periparturient sows housed in crates increases stress and increases the prevalence of savaging behaviour compared to sows housed in loose farrowing environments (Lawrence et al.

1994; Jarvis et al., 1998). Although such negative maternal behaviours are likely to be less evident in extensive systems, where injury does occur from any source there may be more limited opportunities for human intervention to treat such injuries or provide protection. In comparison to intensively managed animals, extensively managed livestock of most farmed species have greater space availability and thus the chances that the newborn will be accidentally trapped, crushed, or stepped on by its own mother or another animal in the same group will be reduced. In the majority of pig housing systems, sows will farrow individually thus risk of piglet injury from other members of the herd is unlikely, however, the risk of crushing by the mother is high and where sows are kept loose (as in extensive systems) crushing is the major source of piglet mortality (Edwards et al. 1994).

Another source of neonatal pain, injury, and death in extensive systems is through predation, which is effectively prevented in intensive systems (as discussed in Chapter 6). The risk of predation is greatest for the small ruminant species, which are frequently managed in areas where large wild carnivores (e.g., bears, wolves, lynx, coyotes, mountain lions) are also common, and lambs and kids are also vulnerable to avian predators. In comparison to adult animals, lambs or kids are more likely to be selected by predators as the target for attack, and less likely to survive an attack than an adult. In addition, ewes and does have very limited abilities to drive off predators and protect their young from attacks, because they themselves are also vulnerable to predation. Maternal cows are more able to provide some physical protection of their calves, and cows with calves at foot can be very aggressive to perceived predators. Although the nest offers some protection, neonatal piglets are vulnerable to opportunistic carnivores, such as badgers and foxes, which have been observed taking newborns out of huts during farrowing when the sow is at her most somnolent and is unable to be vigilant. In addition, more than half of outdoor pig producers in the UK report some losses of piglets to wild predators.

Unlike humans and rodents, cattle, sheep, and pigs do not transfer maternal antibodies across the placenta to the foetus, thus the neonate is born vulnerable to various infectious diseases (such as *E. coli* infection known as "watery mouth disease" in lambs), until they can obtain adequate passive immunity through sucking colostrum from their mothers (Hodgson, 1994; Hodgson et al., 1995). The newborn gut is initially permeable to the passage of macromolecules, in order to facilitate the passage of immunoglobulins into the neonatal blood stream. However, this does also leave the neonate vulnerable to ingesting pathogens from the environment which can also pass through the gut wall. The permeability of the gut gradually decreases over time, and colostrum ingestion itself accelerates the process of gut closure (Sangild et al., 2000), thus preventing this route of neonatal infection. This further emphasises the importance of early sucking as this provides protection against infectious disease, and delayed sucking can both reduce the passive transfer of immunity (as the gut closes), and increase the risk of infection through ingestion of pathogens. In intensive systems, neonates may be more exposed to pathogens than in extensive systems, but human intervention to provide antibiotics is also a possibility which may occur to a lesser extent in extensive systems.

The risks of hunger, thirst, and cold are greater in extensive systems than in intensive systems, although whether management affects the risks of experiencing pain, injury, and disease is not clear. However, in extensive systems, there is a greater opportunity to express species-typical behaviours. The next sections will consider the specific issues around the expression of those behaviours and neonatal mortality in the major farmed species that are commonly managed extensively. To provide some comparative context for the different species, examples of average mortality rates for each species are given in Table 8.1.

NEONATAL MORTALITY IN EXTENSIVELY MANAGED SHEEP

Estimates of lamb pre-weaning mortality in domestic sheep range from 10% to 30% (Dwyer, 2008b; Table 8.1), although the lower figure tends to be associated with housed or intensively managed animals. Generally an average figure of 15% mortality is common across all sheep-producing countries, although considerable between-flock variation is known to occur. In many of the largest sheep producing countries (Australia, New Zealand, UK), most ewes are managed to lamb outside in extensive conditions. The main causes of lamb mortality are related to trauma experienced during the birth process, failure of neonatal adaptation to postnatal life (e.g., inability to maintain body temperature, low lamb

Table 8.1 Comparative published pre-weaning mortality rates of extensively managed farm livestock species. Some studies do not distinguish between stillbirths and neonatal mortality but the figures reflect the overall losses of young livestock.

Species	Mortality (%)	Country/comments	Reference
Sheep	11.8	UK hill flock	Sawalha et al., 2006
	5.4–24.4	New Zealand studies	Fisher, 2003
	10.9–29.8	Western Australia	Kelly, 1992
Beef cattle	6.7	Range cattle, USA	Patterson et al., 1987
	9.5	Australian tropical breeds	Bunter et al., 2014
Pigs	12.0	Outdoor farrowing arcs, UK	Kilbride et al., 2012
	12.0–17.9	Outdoor farrowing huts, UK	Baxter et al., 2011a
	15.1	Outdoor farrowing huts, Sweden	Wallenbeck et al., 2008
	11.4–19.7	Outdoor farrowing huts, USA	McGlone & Hicks, 2000
Camels	30.0	To 90 days, Ethiopia	Ahmed & Hegde, 2008
	30–59	Kenya	Mukasa-Mugerwa, 1981
	21.9–27.1	Kenya	Kaufmann, 2003
Goats	17.4	Up to 6 months, India	Singh et al., 2014
	12.6	Ethiopia	Deribe et al., 2014
	37.0	Brazil	Soares et al., 2010
	21.5	Mexico	Mellado et al., 1991
Deer	10.0	New Zealand	Wass et al., 2003
	11.6	UK	Blaxter & Hamilton, 1980

vigour, poor establishment of a maternal bond), infectious disease, functional disorders, and predation. The compound effects of starvation, mismothering, and hypothermia are a major contributor to mortalities in extensive management systems, acting as either a direct or indirect (e.g., through weakness and vulnerability cause) of death. In this section the behaviours and biology of the ewe and lamb around birth, which contribute to lamb survivability, will be considered.

The sheep is highly social and breeds seasonally (except in tropical regions) and synchronously. Thus all the ewes within the flock conceive and carry their lambs at the same time, generally through the winter months, and give birth together within a short period of time in the spring. This ensures that lambs are born when spring grass is plentiful and the ewe will be able to produce the most milk. Breeding synchronously means that all the neonatal lambs are present in the flock at the same time, which reduces the threat of predation on each individual lamb. However, synchronous breeding also means that ewes and lambs need a mechanism to reliably recognise one another within a mobile flock made up of many mothers and offspring. Wild sheep, or more primitive breeds, usually produce a single lamb for each pregnancy but selection and domestication (coupled with improved nutrition around conception) has led to an increase in litter sizes such that twin and triplet lambs are common in some systems, and four or even five lambs per litter are also possible. In extensively managed systems, however, it is often desirable that the ewe produces no more than two lambs, which she should then be able to rear unaided.

Although the ewe is normally very gregarious, as she reaches the end of pregnancy she may move away from the flock and select a birth site that is isolated from the social group. Wild sheep, such as the Mountain Bighorn in North America, are reported to spend up to two weeks away from the social group at lambing, but domestic sheep may spend only a few hours or days away from the flock (Dwyer and Lawrence, 2005). Important features of the birth site include shelter, absence of wind chill, a dry bed, and a south-facing aspect. Thus, the ability of ewes to select an appropriate birth site, and the availability of birth sites with the desired characteristics in the extensive environment, may help to protect the neonatal lamb from hypothermia. In addition to the protection that might be afforded to the lamb by the birth site, maternal social isolation can aid the ewe in developing a bond to her neonatal lamb without the interference of other ewes. The importance of the birth site for ewe–lamb bonding has been demonstrated by the increased survival of twin lambs with an increase in time spent at the birth site by their mother (Putu et al., 1988). However, the ability of the ewe to show isolation behaviour and birth site selection is much reduced in intensive systems, and movement of the ewe from her preferred site, or interference by other ewes, is greater in housed animals.

The first signs of imminent birth in the ewe are an increase in restlessness, circling and pawing the ground, and licking the lips. The ewe also shows intense interest in any amniotic fluid that may be spilt on the ground, which prior to the onset of labour has been avoided. Birth is usually a relatively quick process, with the lamb being expelled approximately 30 minutes after the appearance of the amniotic fluids at the vulva of the ewe. As a prolonged labour can lead to brain trauma and hypoxia in the neonate and impairs sucking, locomotor activity, and thermoregulation, a swift labour is important for lamb survival. In extensively managed flocks, often obstetric assistance for ewes experiencing birth difficulty, or dystocia, may not be available or may be delayed, and this can affect not only the lamb but may also be a cause of maternal mortality. However, sheep breeds that are normally managed to give birth outdoors, have been shown to have quicker and easier deliveries than breeds of ewe that may be more commonly lambed indoors (Dwyer and Lawrence, 2005).

Prior to birth most ewes are disinterested in, and may even be aggressive towards, newborn lambs. The process of parturition alters ewe responsiveness to produce an intense and focused interest in her lamb. This is expressed by rapid licking or "grooming" behaviour (Figure 8.1), starting at the head of the lamb and then working along the whole body, accompanied by many low-pitched, throaty bleats or "rumbles." Ewe licking behaviour serves to dry and stimulate the lamb, and helps to clear any residual amniotic fluid and membranes from the

Figure 8.1a-b: a, Licking or grooming behaviour performed by the parturient ewe for approximately 4 hours after birth (photo Steven Johnstone); **b,** teat-seeking behaviour by the newborn lamb (photo Ann McLaren)

mouth and nose of the lamb to aid the onset of breathing. However, licking and the smell and taste of the lamb also play a vital role in ewe bonding to the lamb. Initially the ewe is attracted to the smell of amniotic fluids on the coat of any newborn lamb. Then she learns the smell and taste of her own lambs to form an "olfactory memory" for her lamb in the first hour after birth. After this she becomes "selective" for her own lambs and will restrict her maternal care to her own offspring only and will reject any attempts to suck by lambs that are not hers (Poindron et al., 2007). The period when the ewe will form this selective attachment is rather short after birth, and ewes that are unable to smell or lick their lambs during this period may fail to recognise them, or cease to show maternal behaviour at all. If the ewe does not learn to make this association with her lamb it is very likely that the lamb will die as any other maternal ewe will be selective only for her own offspring, and will not let the un-mothered lamb suckle. This mechanism allows the ewe to form an exclusive attachment to her own lamb, and ensure that her milk supply is available only to her own offspring. Such failures in the formation of the ewe–lamb bond may occur if more than one ewe is lambing in close proximity and ewes become attracted to lambs that are not their own, sometimes in the later stages of labour before they give birth themselves. The mismothering that can then occur may leave a single ewe having formed a bond to more lambs than she can properly provide milk for, and some lambs without a mother. In extensive systems, where the ewe is able to show isolation responses, it may be comparatively rare that she becomes separated from her offspring during the critical period, but without human intervention to feed any abandoned lambs they will be particularly vulnerable.

Although the maternal behaviour of the ewe is extremely important for lamb survival, the lamb is not a passive player in the bonding process. Lambs are precocious (so well developed at birth), and show rapid behavioural development to stand and seek the udder within a few minutes of being born (Figure 8.1). Although ewe behaviour can help to facilitate the teat-seeking behaviour of the lamb, for example, by standing still, crouching, and turning out a hind-leg to make the udder more accessible, the lamb needs to be sufficiently active to stand, orient towards the ewe, integrate the sources of sensory information from the ewe (such as the differences in texture between the woolly belly of the ewe and the smooth feel of the udder, the smell of the wax produced by the udder), and attach to a teat. As lambs are born with finite body reserves, which can be rapidly depleted particularly in cold weather, they need to accomplish these behaviours as soon as possible after birth. Lambs that stand and suck quickly after birth (within an hour of delivery) have frequently been shown to have better survival than lambs that are slow to perform these behaviours (Dwyer and Lawrence, 2005). Over time the unsuckled newborn lamb becomes weaker and less vigorous in its attempt to find the udder and will gradually succumb to hypothermia and starvation.

The sucking behaviour of the lamb, whilst vital to ensure that the lamb is properly nourished, also plays another important role in the development of the ewe–lamb bond: as the ewe needs to learn to recognise and distinguish her lamb from all others, so the lamb also needs to be able to identify its own mother. Sheep are a "follower" species (where the offspring follow their mothers closely from birth, unlike hider species [see below]; Lent 1974), thus the lamb needs to keep up with its dam soon after birth. This requires that the lamb not only shows appropriate locomotor competence, but that it can discriminate its own dam from other ewes and is attracted to remain with her. Studies in Merinos suggested that separation between ewe and lamb is a major contributory factor to lamb mortality (Stevens et al., 1982; Alexander et al., 1983). Lambs are initially attracted to any large objects, but can recognise their mothers at close quarters as early as 12 hours after birth (Nowak et al., 1987), and at a distance by 24 hours old (Nowak, 1991). Recognition by the lamb begins through the act of sucking from the ewe – the first sucking interactions with the ewe are strongly rewarding and play an important role in establishing a preference in the lamb for its own mother (Nowak and Poindron, 2006). Vocal communication between ewe and lamb is also important to maintain the relationships, and aid both partners to become reunited if they have become temporarily separated.

There are a number of external and internal factors that can disturb the normal process of the development of the ewe–lamb relationship and influence the risk of lambs dying, including the weather, maternal experience, maternal nutrition, and the genotype of ewe and lamb. In extensive systems exposure to environmental and climatic variables play a major role in lamb survival. Lambs have brown fat reserves located around the major organs and at the scapular, which possess a unique uncoupling protein that allows the lamb to rapidly generate large amounts of heat by a process called non-shivering thermogenesis. This can provide the lamb with sufficient heat to maintain body temperature immediately after birth, but this requires energy to keep up the levels of heat production. The newborn lamb needs to replenish those reserves by suckling, so in cold temperatures (particularly if it is also windy or wet), where lambs are rapidly losing heat to the environment, the interval in which the lambs can maintain body temperature before suckling will be very short. Survival under these conditions is then critically dependent on the rapid standing and sucking by the lamb, and any factor that extends this interval will increase the risk of the lamb dying.

Maternal inexperience, in particular ewes lambing for the first time, is associated with less competent onset of maternal behaviour, reduced milk production by the ewe, and an increase in lamb mortality (Dwyer, 2008b). Inexperienced ewes frequently express fear-like behavioural responses towards their lambs initially, and are less likely to stand still after birth to allow their lambs to suck, resulting in prolonged intervals to first sucking in the lamb. With time, however, and experience of interacting with their lamb, these mothers can learn to show a high quality of maternal care. In extensive systems the offspring of primiparous ewes (those lambing for the first time) are likely to be more vulnerable to any additional factors, such as inclement weather, than the offspring of more experienced ewes where lambs will suck more quickly.

Nutrition of the pregnant ewe, especially in late gestation, is harder to manage in extensive situations, compared to intensive, where ewes get most or all of their nutritional intake from grazing. Cold temperatures, drought, altitude, and underlying soils can all have an impact on the quality and quantity of grazing available to the ewes, and management factors will influence whether supplementary nutrition is provided or not. Maternal nutrition affects the prenatal growth of the lamb, affecting the development of brown fat reserves (Ojha et al., 2013) and body weight (Dwyer, 2008b). Low birth weight lambs have a greater surface area to body mass ratio than larger lambs, and so lose more body heat to the environment. They are also less able to generate heat due to their reduced amount of brown fat, and are less vigorous (Dwyer, 2003) so are slower to reach the udder and suck. In combination these factors increase the likelihood that the low birth weight lamb will not survive. As well as affecting the behaviour of the lamb, low nutrition also reduces the expression of maternal behaviour in the ewe (Dwyer et al., 2003), and affects her milk production. Poor nutrition of the late pregnant ewe, which is more likely to occur in extensive systems, is therefore a major risk factor for lamb mortality.

Generally the breeds of sheep that commonly lamb outdoors have more rapid lamb behavioural development, and express a greater quantity and quality of maternal care than those that are normally managed to lamb indoors (Dwyer and Lawrence, 2005). Extensively managed sheep are still subjected to a degree of natural selection, which has likely ensured that active lamb behaviours and focused maternal responses are selected for and retained in the breeding population. The existence of breed differences in maternal behaviour suggest that these might be under some degree of genetic

control, and selection for ewes on ability to rear lambs is possible. However, no studies have specifically estimated genetic parameters for maternal behaviours in sheep, and most studies suggest a poor genetic correlation between maternal behaviour and lamb survival (Brien et al., 2014). Selection on lamb traits has so far proved a more fruitful route to improve lamb survival. Although the heritability (the proportion of the variance in a trait that is related to the genetic background of the animal) of lamb survival is very low and close to zero, there is good evidence that indirect measures of lamb survivability have a reasonable heritability. Assessment of the degree of birth difficulty, and the vigour and sucking ability of the lamb, have been shown to be related to lamb survival and to have good heritability (Matheson et al., 2012). Neonatal rectal temperature and latency to bleat following handling have also been shown to be moderately genetically associated with lamb survival (Brien et al., 2014). These data suggest that it may be possible to improve lamb survival in extensive systems by selecting for lamb behavioural and thermoregulatory traits. However, as mentioned above, and elsewhere, extensive management systems still require some human intervention, and increasing the survival capacities of lambs does not mitigate any requirement for supervision or responsibility for their welfare.

NEONATAL MORTALITY IN EXTENSIVELY MANAGED BEEF CATTLE

In comparison to other species (Table 8.1) beef cattle generally have lower mortality rates at less than 10%. However, as this species most commonly delivers a single calf, the relative production losses with neonatal calf mortality may well be greater: often the ewe or sow that loses a neonate still remains a productive animal, thus some of the investment in her pregnancy can still be recouped, whereas the beef cow that loses a calf is usually no longer productive in that pregnancy. Welfare issues relating to calf death are, however, similar to other species. As with sheep, calf perinatal deaths can be attributed to birth difficulty, failure to adjust to extra-uterine life (generally associated with weakness or low birth weight), and infectious disease (becoming more important as a cause of mortality beyond the immediate postnatal period). Calving difficulty or dystocia is consistently found to be related to high calf mortality occurring within 24 hours of birth, with mortality increasing with the severity of the dystocia (Nix et al., 1998). Nearly half of all calf mortality in first parity heifers, and a quarter of all calf mortalities in cows, are associated with dystocia (Eriksson et al., 2004), demonstrating the importance of this factor in calf mortality.

Cows are a social species, living in matrilineal groups. However, as birth becomes imminent mothers separate themselves from the herd and seek isolation to calve. This isolation-seeking response is more pronounced in animals living in wooded or forest areas, but still occurs to a lesser extent in open grasslands (Lidfors et al., 1994). Cows generally choose to calve in shelters if available, or choose elevated areas, with a dry and soft surface and vegetation to provide cover. Social isolation ensures that mothers are able to form close attachment to their own offspring, without interference or cross-licking from other cows. Thus provision of suitable calving sites in extensive systems will help to increase the chance that cows are able to bond successfully with their own offspring so improving the likelihood that calves will survive. Providing cows with suitable places to calve in the extensive environment may also help to reduce the incidence of birth difficulty which is a significant cause of neonatal mortality in this species. This may be particularly important for heifers, which have both higher rates of dystocia than cows (Dargatz et al., 2004), and higher offspring mortality.

The onset of maternal care in cattle is similar to that of sheep, with a short period of high-intensity licking occurring immediately after birth. Although this has been much less studied than in sheep, it is likely that this behaviour performs a similar function in cattle, aiding in cleaning, drying, and stimulating the calf, and playing a role in the development of a preferential relationship between cow and calf. There is less evidence, however, for the formation of such an exclusive attachment in cattle as is seen in sheep, and cross-suckling (where calves suckle from cows that are not their mother) is comparatively common, particularly in twin calves. Mother cows continue to lick their calves frequently over the first six or more months of life, although this does decline with increasing calf age, and has been suggested to help re-establish and maintain the bonds between mother and young. Following delivery, calves stand within an hour of birth, and first sucking, which can be a prolonged bout, occurs within two hours of birth. Many of the risk factors for calf mortality are similar to those for low calf vigour: cow age, calf gender, the incidence of dystocia, ambient temperature on the day of birth, low body condition in the dam, low birth weight, and breed all contribute to the incidence of low vigour calves (Ogata et al., 1999; Riley et al., 2004). Very young and very old (more than 11 years) cows tend to have a higher incidence of low calf vigour. Young dams are still growing themselves, therefore they need to partition nutrients to sustain their own growth as well as that of their foetus. Thus they may be more likely to produce small calves that are more susceptible to low vigour. Low calf vigour in older cows may be due to an impaired ability to match nutrient requirements to sustain lactation and body maintenance requirements, resulting in a reduced colostrum and milk intake by the calf. In addition, older cows may have poorer udder and teat conformation which can affect the ability of the calf to find the teat and suck quickly after birth.

As with other species male calves appear to be less vigorous than females at birth, and have consequently lower survival, and this reduced vigour at birth appears to be in addition to the increased likelihood of dystocia (Riley et al., 2004). The reasons for low vigour in males are not clear, however, males require greater maternal investment before birth, and may have more difficulty adjusting to postnatal life than females. Low birth weight calves, and males, are reported to receive more protective maternal care and more frequent sucking bouts than heavier calves or females respectively (Stehulova et al., 2013). This may be related to more frequent solicitation of sucking bouts in these animals, which may be related to hunger and increased need for maternal milk. Studies suggest that there are breed and sire effects on the incidence of low calf vigour (Ogata et al., 1999; Riley et al., 2004) and heritability of both calf vigour and calving ease is relatively low but significantly greater than zero. Thus selection for improved calf vigour at birth is possible, although improved management to reduce low vigour will have a more immediate impact on calf survival. Of particular relevance to extensive management is the influence of low ambient temperatures in reducing neonatal calf vigour, although rainfall does not appear to have an effect. As cows do seem to calve in shelters when these are provided, this can improve the survival of calves by protecting them from the effects of cold weather.

Unlike sheep (which are "followers"), cattle are considered to be somewhat intermediate between a "hider" species (such as deer; Lent, 1974) and a follower. Evidence suggests that they are flexible in their behavioural strategies and able to adopt either strategy, depending on topography (e.g., hiding seems to be more common in forested areas) or familiarity with terrain (e.g., hiding has been reported in studies with breeds of cattle that have been maintained in the same environment for many generations). In these studies the calves may remain at the birth site for a number of days, before the cow re-joins the herd with her calf, whereas in other studies the calf may move from the birth site on the day of birth (Lidfors and Jensen, 1998). The suckling pattern in cows may, however, be more similar to that of a hider species as suckling occurs mainly early in the day and in late afternoon. Typically calves are fed in four or five suckling bouts per day, each lasting up to

10 minutes, unlike a true follower such as the sheep where lambs are fed in frequent, shorter sucking bouts throughout the day. Cattle also perform "crèching" behaviour, where calves preferentially group together, often to rest or play, forming a separate social group from the adult cows. This may be facilitated by an increasing preference for maternal cows to group together, and apart from pregnant animals, once they have calved, such that calves are also in closer proximity to one another. These calf social groups are seen to be formed from about 6 weeks of age, with calves choosing to associate with others of a similar age to themselves, forming bands of up to 25 calves. In some situations "guard" cows are reported, where one or two maternal cows remain in attendance as guardians of the crèche, whilst the remaining cows are able to graze further from the resting calves. Whether guard cows are required or not seems to be related to the size of paddocks (as all cows may be able to maintain visual contact with calves in smaller paddocks) or topographical features.

A particular feature of maternal care in cattle, and of great importance in extensive situations where the mother is largely responsible for the survival of her offspring, is maternal defensive aggression. After calving, mothers show a reduced fear response to novel and potentially threatening situations, and an increase in protective behaviours, especially when their calves are young. This behaviour is considered a component of good maternal care, although whether this behaviour is associated with improved calf survival is not known. However, there is no apparent association between protective maternal care

expressed at birth (e.g., licking the calf and suckling responses) and the level of defensive behaviour that may be displayed to a threatening stimulus (Turner et al., 2013), suggesting that these two components of good maternal care are not linked. The defensive behavioural response is expressed as increased agitation and aggression towards the perceived source of the threat, which may be predators but can also be human stockworkers or passers-by if perceived by the cow to be threats to her calf (Turner and Lawrence, 2007), and declines with calf age. This response can be conflicted in extensive systems, particularly those systems where wild carnivores that are predators of neonatal calves are present and thus where maternal defensive aggression would reduce mortality, but where aggression towards humans can also cause death and injury. Although repeatable within individual cows (i.e., a highly aggressive cow in one pregnancy is likely to be so in a subsequent pregnancy; Hoppe et al., 2008, Turner et al., 2013), the amount of genetic variation in the trait between cows is relatively low in some breeds (Turner et al., 2013). This suggests that defensive aggression is mainly related to management, or other non-genetic factors, that have an influence on the expression of this behaviour, although moderate heritability has been suggested in other breeds (Hoppe et al., 2008). In extensive management systems, with infrequent human contact, cows may not have the opportunity to learn to make more positive associations (such as provision of feed) with humans, and may then treat all humans as if they were predators and a threat to their calves. However, the expression of these behaviours can reduce neonatal losses to predators such as wolves or coyotes which may prey on young calves. Evidence also suggests that, for cattle living in wolf territory, calf mortality may be affected indirectly as calves in herds that have suffered the direct effects of wolf predation have lower average calf body weights (Ramler et al., 2014). This may occur due to alterations in maternal foraging behaviour, to avoid particular areas associated with wolf presence, or as a consequence of maternal stress, but can leave low birth weight calves vulnerable to other causes of neonatal mortality.

Although calf mortality is relatively low, in comparison to other farmed species, it does appear to be rising in some sectors and is a significant welfare and production cost to the beef industry. The main risk factor for calf mortality is dystocia, which can result in low vigour calves, and is particularly a concern in extensive conditions where human supervision and assistance at calving may be infrequent. Calves born following difficult deliveries are slow to ingest colostrum, consequently have lower immunoglobulins in their blood, and are more susceptible to infectious disease. They are also less able to regulate their body temperature than calves born more easily, and may be susceptible to hypothermia. Although the risk of exposure to pathogens may be less in extensive management systems in comparison to intensive systems, this still presents a challenge for neonatal calves. Both calving difficulty and low calf vigour have low but significant heritability suggesting that they could be improved by selection, which would improve calf survival. A more immediate improvement in calf survival could be achieved by improved

management, such as providing shelter in cold weather, to reduce the risk factors that lead to calf mortality.

NEONATAL MORTALITY IN EXTENSIVELY MANAGED PIGS

Although this book has largely focused on herbivores, there are a number of countries, particularly in Europe, that have significant numbers of outdoor herds contributing to the conventional pork supply chain. In the UK, 40% of the breeding herd is extensively managed, in comparison to 5–10% in Germany and France, and less than 2% in the USA. Other systems involving outdoor pig rearing include organic enterprises and pigs produced via the traditional Mediterranean silvopastoral system (e.g., Iberian pigs; Edwards, 2005). Thus we will also briefly discuss the issues surrounding neonatal mortality in extensively managed pigs.

As the pig is a litter-bearing species, some piglet mortality can be considered inevitable (Edwards, 2002). The evolutionary strategy adopted by the sow is to over-produce, thus producing many offspring in the event of good rearing conditions with little investment in piglets that might die early if conditions are unfavourable. The main causes of piglet death in the first 72 hours (when most mortality occurs) are stillbirths, crushing by the sow, and starvation. However, as with all species, these causes of mortality are not always singular, and the primary cause of mortality may be difficult to determine, especially in extensive management conditions where the practicalities of production limit the accuracy of recording the actual number of deaths and the cause. Post-mortem studies suggest that starvation in piglets is lower in outdoor systems compared to indoor (10% vs 15%), but crushing is higher (45% vs 20%; Riart et al., 2000). However, this increase in the proportion of piglet deaths that are caused by crushing does not necessarily mean that mortality is increased in outdoor systems: in the UK outdoor units recorded 16.6% mortality compared to 19.3% in indoor systems.

Sows are social animals, however, about 2–3 days before they are due to give birth they will withdraw from the herd in search of a suitable nest site. Once chosen, nest-building behaviour begins, which includes hollowing out the ground, collecting branches and twigs to border the nest and softer substrates with which to line it. Such highly motivated behaviours are evident in domestic populations of pigs with the outdoor system affording greater opportunities to facilitate these activities than indoor conventional systems (Figure 8.2).

Approximately one hour before the onset of farrowing, sows will enter a quiet phase. Sow behaviour during this phase is characterised by prolonged lateral lying and udder exposure. Despite only rudimentary maternal care during parturition in this species, sows in semi-natural environments will get up during parturition to inspect their offspring, making nose-to-nose contact before rooting the nest to move piglets out of the way and then resuming lateral lying (Jensen, 1986). Such passive behaviour allows piglets to access the udder freely. Newborn piglets are highly motivated (like other neonates) to seek the udder and find a teat. They move along

the udder and sample from each teat eventually settling on a preferred teat to which they will remain faithful and defend from their siblings. Development of this stable "teat order" is important to ensure harmonious sucking bouts leading to maximum colostrum intake and improved survival (Figure 8.3).

In the early stages of lactation, sows generally do not leave the nest site. During this period sows nurse their piglets at 30–70 minute intervals, and mother–offspring bonds are strengthened particularly by rhythmic maternal suckle grunts, which aid offspring in recognising their mother. Post milk-let-down piglets will make nose-to-nose contact with the sow, signalling need and reaffirming bonds and recognition. Such recognition is important for the next stage in lactation which involves leaving the nest and re-joining the herd. Responsibility for staying with their mother at this time is primarily with each individual piglet as sows will only recognise the litter as a whole rather than forming individual attachments.

The risk factors associated with stillbirth and liveborn mortality differ. Risks associated with stillborn mortality are prolonged duration of farrowing, delivery in the last third of the birth order, premature rupture of the umbilical cord, and sow behavioural and physiological characteristics, including parity. These factors often result in fatal hypoxia, or a less viable piglet with poor survival chances post-partum. Sows farrowing outdoors have fewer stillbirths than sows in crated systems (Riart et al., 2000). The likely explanation is that the outdoor system provides many of the stimuli required to complement the natural biology of the pig and facilitate

Figure 8.2a-b-c: Nest-building behaviour in an outdoor system (photos: Marianne Farish)

Figure 8.3a-b: Teat order development in pigs. Piglets will fight to defend their teat. Establishing a stable teat order will promote optimum milk intake and survival. (Photos: Marianne Farish)

Figure 8.3c: (Continued)

good maternal behaviour (Baxter et al., 2011b). In particular it allows full expression of nest-building behaviour. The more complete and functional the nest is, the more likely the sow is to end nest building and begin the quiet farrowing phase (Baxter et al. 2011b) which is likely to influence farrowing duration and therefore the risk of stillbirth. In outdoor managed pigs the nest not only satisfies behavioural needs, it has multiple protective functions for the neonates: as a mattress cushioning sow posture changes and offering a buffer to piglets from fatal crushing and creating a microclimate for the vulnerable newborns. The nest facilitates physical removal of membranes and absorption of birth fluids, reducing the impact of placental fluid evaporation and buffering the piglet from immediate susceptibility to the extra-uterine environment (Baxter et al., 2009).

The piglet is considered the most cold-sensitive ungulate, being born virtually hairless with no brown adipose tissue to facilitate metabolic heat production (Herpin et al., 2002). Birth results in rapid heat loss, affected by physical, behavioural, and environmental factors (Curtis, 1970). Higher air velocity and a larger temperature gradient will also increase heat loss. The nest can help protect piglets from the immediate drop in temperature by slowing the rate of heat loss. However this protection can only do so much, a piglet must rely on behavioural adaptation to gain colostrum and increase its core body temperature. Drinking colostrum increases core body temperature of piglets, and is also important for energy balance and immune protection. Milk production of the sow will be one factor determining how much milk the piglets receive, however, piglet behaviour also plays a role. Piglets will fight to gain teat access and to maintain teat fidelity (Figure 8.3) and, if they are unable to perform optimal udder massaging and suckling behaviours, teat function may be impaired. Piglets failing to establish teat fidelity grow more slowly (De Passillé et al., 1988) and have to get by on opportunistic suckling or they will starve. However, even if a piglet possesses good vigour, a large litter size, where piglets outnumber functional teats, will require managerial intervention to prevent starvation (Baxter et al., 2013) and such interventions are not easy in an outdoor system.

In order for outdoor systems to be a success the genotype of sows must be considered carefully, and well-managed systems rely on selecting for good maternal behaviour, and considered management of farrowing huts and substrate provision to ensure a suitable nest site, facilitating nest-building behaviour and a protective micro-climate. If these environmental strategies are in place further improvements to piglet survival are limited to nutritional intervention to improve embryo quality, subsequent birth weight and uniformity of piglets (e.g., fermentable fibres and essential fatty acid supplementation in pregnancy diets; Van den Brand et al., 2009; Rooke et al., 2001) and genetic selection strategies. This latter tactic has had notable success in outdoor systems; emphasis within breeding indices was placed on numbers weaned rather than numbers born to improve survival rates (Roehe et al. 2010). Breeding for

improved maternal behaviours is likely to reduce neonatal losses even further (Baxter et al. 2011a).

NEONATAL MORTALITY IN OTHER EXTENSIVELY MANAGED SPECIES

Goats and camels, along with sheep, are important species in the developing world, where they are managed as part of traditional or nomadic agropastoralism. The particular issues associated with neonatal mortality in these species will be considered here, as will mortality in deer, which have only recently been domesticated and managed.

Goats

Studies considering neonatal kid mortality in South Africa (communal goat farming conditions), India, Turkey, and Brazil suggest that one of the main causes of mortality is disease. Neonatal infections, and parasites including ectoparasites or ticks, are the most commonly cited diseases leading to the deaths of neonatal kids. In some studies mortality is considered to be lower in extensive systems than in intensive systems, perhaps because the lower flock sizes in extensive systems reduce the spread of pathogens. However, in other reports mortality is higher in extensive systems, with poor quality nutrition in marginal grazing lands identified as the main risk factor, leading to poor colostrum production and increased susceptibility to disease.

Goats have broadly similar behavioural responses to sheep in the neonatal period and kids also need to be active and vigorous to stand and suckle adequate colostrum to provide protection against infectious disease. In nomadic or communal grazing systems often the pastures may be highly variable and there is little ability of herders to control pasture quality beyond selecting where they can take their flocks to graze. Low birth weight and low vigour kids, combined with poor quality colostrum, which can both be attributed to poor maternal nutrition during pregnancy and lactation, are significant risk factors for impaired transfer of colostrum. In systems where sheep and goats are kept as mixed species flocks under the same management, goat kids are considered to have higher mortality than lambs, although the reasons why this might be so is not clear. However, newborn kids are particularly susceptible to cold stress (Mellado et al., 2000) and highest kid mortality occurs in the cold season in several countries. In contrast, although birth weights of kids from goats exposed to high air temperatures during pregnancy are somewhat reduced, lowest goat kid mortality occurs in hot and humid conditions, and kids seem to be relatively resistant to high temperatures.

Camels

Camel calves appear to have particularly high mortality in comparison to other species (Table 8.1). The very highest figures (up to 50% mortality) have been considered an overestimation by Kaufmann (2003), but even revised figures of up to 30% are greater than for other extensively managed species. About 50% of mortality occurs in the first week of life. The main causes of mortality in camel calves are considered to be disease, predation, trypanosomiasis, the consequences of drought, and browsing of poisonous plants (in older calves).

Camels are kept generally by nomadic pastoralists, particularly in Africa, for transport or draught purposes and milk production; the meat is rarely eaten except in exceptional circumstances. Generally they produce under harsh environmental conditions: high temperatures, water shortages, and low food quality and quantity. Camels may form part of mixed species herds, with small ruminants and cattle, which are mobile so enabling movement to better forage availability. Within these systems, breeding females have high value and, because of their relative longevity (more than two decades), are considered important members of the pastoralist's "family." Camels are a slow maturing species, with a gestation of almost 12 months, and may not calve for the first time until they are between 4 and 6 years of age (Kaufmann, 2005) compared to 1 to 2 years in other livestock species, and at a calving interval of between two and three years. The high calf mortality is therefore a particular issue following the long investment in the growth and development of the mother.

Studies of the parturition process of camels (reported in Mukasa-Mugerwa, 1981) suggest that separation and social isolation at birth in camels occurs only infrequently, although whether this is due to the mobile nature of camel herds, or biological needs of the camel is not clear. Birth is relatively quick, following a period of restlessness and agitation, but, unlike goats, sheep, or cattle, the camel does not usually lick her calf. Camel herders may intervene to dry amniotic fluids from the calf, stretch the limbs of the calf, place it in front of its mother, and move it to the shade if born in direct sunlight. However, although some herders are aware of the importance of colostrum in protecting against disease, there is also a belief that first colostrum is dangerous for calves – it is considered too strong and a cause of diarrhoea (Kaufmann, 2003). Camel calves may be deliberately prevented from suckling for a period, and their ingestion of colostrum can be restricted and very low. Camels have been found to produce a high concentration of immunoglobulins in colostrum, but calves had critically low levels suggesting that poor ingestion of colostrum is the main issue in the low immunoglobulin status of calves (Kamber et al., 2001). These data suggest that part of the high mortality of camel calves from disease is probably related to low passive transfer of immunity via inadequate colostrum intake. In times of drought, calves also need to compete with humans for access to the milk supply of their mothers, which may lead to weakened calves, and perhaps an increase in browsing on unsuitable vegetation.

Red Deer

Cervids have only recently been domesticated and farmed, in comparison to the other species considered in this chapter, and are usually farmed under extensive management. Neonatal mortality rates are rather similar to those reported in sheep and cattle (Table 8.1), and the main cause of mortality is considered to be low calf birth weight (Blaxter and Hamilton, 1980), and disturbances or inadequate environmental conditions that can affect maternal responses and suckling in hinds (Birtles et al., 1998). Like pigs, deer

appear to have a stronger motivation to isolate themselves as birth approaches in comparison to other ruminant species, and pacing along the fence lines in enclosed paddocks occurs about 24 hours prior to birth, although adult hinds may isolate themselves up to two days before giving birth (Wass et al., 2003). In farming systems, however, opportunities to be properly isolated might be limited and interference by other animals is common, occurring in half of all births, particularly in the births of inexperienced yearlings. Deer may, as described for cattle, adopt different isolation and parturition strategies depending on the nature of the environment, particularly the presence of vegetation cover, and perceived predation risk (which may be the risks of disturbance by humans).

In the wild, red deer move to higher ground to calve, and may stay away from the matriarchal herd for two weeks (Clutton-Brock and Guinness, 1974). Immediately after birth, hinds remain close to their calves, and they are vigorously licked as previously described for domestic sheep and cattle. Thereafter, hinds will leave the calves for longer periods, and may be grazing up to a kilometre away, returning two to four times a day to feed their calves, before re-joining the herd when the calves are 2–3 weeks of age. Clearly, in a managed system, it is difficult to provide opportunities for hinds to express these ranging and isolation behaviours at parturition. It is likely that the inability to express these behaviours will contribute to calf mortality due to increased anxiety or frustration (as expressed by the fence line pacing behaviours). In addition, the high rate of interference by other hinds with the birth process, particularly of yearlings, may contribute to the higher mortality of their offspring in comparison to older animals. However, in common with other ungulates, the calves of yearlings are generally lighter and slower to stand and suck than calves of adult hinds, which may further contribute to their lower survivability.

Mechanisms to increase the survival of the calves could include providing more opportunities for hinds to isolate themselves, better vegetation as cover to give more preferred birth sites, and reduced disturbances by human interactions with deer. However, many of these interventions might be impractical or difficult to implement in a farming system, given the very large ranges and distances covered by deer in the wild. An alternative strategy might be to select deer for increased social tolerance, as suggested by Pollard (2003), which could reduce the impact of interferences in the birth processes of other hinds.

OPTIONS TO IMPROVE NEONATAL SURVIVAL AND ENHANCE ANIMAL WELFARE: WHAT ELSE DO WE NEED TO KNOW?

In the different species described here that are managed extensively, the main causes of mortality differ somewhat in their prevalence, but a key feature in all is the reliance on appropriate maternal care and offspring behavioural responses for neonatal survival. In all species the probability of neonatal survival is improved when the neonate reaches the udder quickly and is able to ingest sufficient colostrum to provide adequate nutrition, fuel for thermoregulation, and immunological protection. This in turn is dependent on species-appropriate maternal behaviours, sufficient colostrum production by the mother, and, since all these species are precocious (or semi-precocious in the case of the piglet), sufficient vigour from the newborn to locate the udder and attach to a teat. The most important options for improving survival, therefore, are those management or genetic approaches which will enhance expression of these behaviours.

In all the species described in this chapter, the place in which the mother might choose to give birth plays a vital role in her comfort and security at a time when she is particularly vulnerable, and serves to ensure that appropriate maternal behaviours are expressed. Understanding the importance of the birth site, and providing the right environmental conditions for parturient livestock, are clearly very important in improving the chances of offspring survival by enhancing the expression of maternal care. The place where offspring are born will also play a role in protecting the neonate, from predators or from climatic extremes, so providing the right conditions for mothers, whether that be nesting substrates for pigs or vegetation cover for hinds, may ensure that offspring are born in the most appropriate place.

In cattle, sheep, and pigs, studies have already been conducted to assess the possibility of enhancing the expression of desirable maternal and neonatal behaviours by genetic selection. These survival traits are typically difficult to measure, measurement may be very imprecise, and heritability is generally low (Brien et al., 2014). However, where more focused measures, for example of specific components such as birth difficulty, piglet survival, or lamb vigour, have been assessed as offspring traits then significant genetic contributions to these traits have been identified (Roehe et al., 2009; Matheson et al., 2012;). It is possible, therefore, that appropriate behaviours to improve survival could be targets for genetic selection. An alternative approach for these difficult-to-measure traits, is to use genomics or gene markers. Marker-assisted selection relies on identifying a small region of DNA which contains an important gene for the expression of a particular trait. However, with complex traits like behaviour, this may be very difficult as it is likely that many genes are important and thus many markers might be required. Genomic selection covers the whole genome and thus can capture all the genetic variation. Although not yet widely used, difficult-to-measure traits and those that can only be measured later in life, such as maternal care, are ideal candidates for genomic selection and may offer the potential in the future to produce animals with enhanced survival capabilities.

Extensive systems, particularly those where animals may graze unfenced pastures, may appear to be akin to allowing animals to be virtually wild. However, these systems may still be highly managed, even if there is considerably less direct human–animal contact, and stockperson decisions will impact on the neonatal survival of offspring. This may be related to the provision of appropriate nutrition at key stages in the gestation and rearing of offspring, environmental and health management, as well as what limited human intervention may be possible or necessary in different systems. Indeed, in species that are particularly sensitive to disturbance, appropriate stockperson behaviour may be to minimise the interventions by humans around parturition to allow the birth process and initial onset of mother–offspring bonding to occur without interruption. Thus, as in other systems, the training, knowledge, and skills of the stockperson will be crucial in influencing the welfare of animals and the survival of newborn livestock.

CONCLUSIONS

As shown in Table 8.1, the mortality of extensively managed livestock can be high, and impacts on the welfare of animals in these systems, and the profitability of farms relying on production from extensively managed livestock. However, few studies have directly compared intensive with extensive systems, and in some cases the survival of extensively managed livestock may be better than for those in more confined systems. With increasing pressures on global land use from a burgeoning human population, and increased demand for human food, extensive animal production is being pushed to more marginal and poorer quality lands, where only animals that are efficient at using grass can survive and reproduce efficiently. Extensively-managed mothers need to be able to cope with poor quality nutrition but still provide for the prenatal growth of their offspring and the energetic demands of lactation. Newborns in these systems need to be able to cope with the environmental and climatic extremes they may face even on the day of birth. As these systems are low input, and human intervention around birth can be virtually impossible, the mother and her offspring need to be self-sufficient, and able to use their full range of behavioural and physiological adaptations to survive. Understanding what the mother and offspring need to express these adaptations, and so providing an appropriate environment throughout pregnancy, birth, and lactation, will give these mechanisms the best chance to enhance neonatal survival.

REFERENCES

Ahmed, S.M., and Hegde, B.P. (2008) Preliminary studies on the major important camel calf diseases and other factors causing calf mortality in the Somali Regional state of Ethiopia. In: Gahlot, T.K. (Ed.) *Proceedings of the International Camel Conference "Recent Trends in Camelids Research and Future Strategies for saving Camels", Rajasthan, India, 16-17 February 2007.* pp. 31–41.

Alexander, G., Stevens, D., Kilgour, R., de Langen, H., Mottershead, B.E., and Lynch, J.J. (1983) Separation of ewes from twin lambs: Incidence in several sheep breeds. *Applied Animal Ethology* 10, 301–317.

Algers, B., and Uvnäs-Moberg, K. (2007) Maternal behavior in pigs. *Hormones and Behavior* 52, 78–85.

Barrier, A.C., Ruelle, E., Haskell, M.H., and Dwyer, C.M. (2012) Effect of a difficult calving on the vigour of the calf, the onset of maternal behaviour, and some behavioural indicators of pain in the dam. *Preventive Veterinary Medicine* 103, 248–256.

Barrier, A.C., Mason, C., Dwyer, C.M., Haskell, M.J., and Macrae, A.I. (2013a) Stillbirth in dairy calves is influenced independently by dystocia and body shape. *The Veterinary Journal* 197, 220–223.

Barrier, A.C., Haskell, M.J., Birch, S., Bagnall, A., Bell, D., Dickinson, J., Macrae, A.I., and

Dwyer, C.M. (2013b) The impact of dystocia on dairy calf health, welfare, performance and survival. *The Veterinary Journal* 195, 86–90.

Baxter, E.M., Jarvis, S., D'Eath, R.B., Ross, D.W., Robson, S.K., Farish, M., Nevison, I.M., Lawrence, A.B., and Edwards, S.A. (2008) Investigating the behavioural and physiological indicators of neonatal survival in pigs. *Theriogenology* 69, 773–783.

Baxter, E.M., Jarvis, S., Sherwood, L., Robson, S.K., Ormandy, E., Farish, M., Smurthwaite, K.M., Roehe, R., Lawrence, A.B., and Edwards, S.A. (2009) Indicators of piglet survival in an outdoor farrowing system. *Livestock Science* 124, 266–276.

Baxter, E.M., Jarvis, S., Sherwood, L., Farish, M., Roehe, R., Lawrence, A.B., and Edwards, S.A. (2011a) Genetic and environmental effects on piglet survival and maternal behaviour of the farrowing sow. *Applied Animal Behaviour Science* 130, 28–41.

Baxter, E.M., Lawrence, A.B., and Edwards, S.A. (2011b) Alternative farrowing systems: Design criteria for farrowing systems based on the biological needs of sows and piglets. *Animal* 5, 580–600.

Baxter, E.M., Rutherford, K.M.D., D'Eath, R.B., Arnott, G., Turner, S.P., Sandoe, P., Moustsen, V.A., Thorup, F., Edwards, S.A., and Lawrence, A.B. (2013) The welfare implications of large litter size in the domestic pig II: Management factors. *Animal Welfare* 22, 219–238.

Bellows, R.A., and Lammoglia, M.A. (2000). Effects of severity of dystocia on cold tolerance and serum concentrations of glucose and cortisol in neonatal beef calves. *Theriogenology* 53, 803–813.

Birtles, T., Goldspink, C.R., Gibson, S., and Holland, R.K. (1998) Calf site selection by red deer (*Cervus elaphus*) from three contrasting habitats in northwest England: Implications for welfare and management. *Animal Welfare* 7, 427–443.

Blaxter, K.L., and Hamilton, W.J. (1980) Reproduction in farmed red deer. 2. Calf growth and mortality. *Journal of Agricultural Science* 95, 275–284.

Brien, F.D., Cloete, S.W.P., Fogarty, N.M., Greeff, J.C., Hebart, M.L., Hiendleder, S., Edwards, J.E.H., Kelly, J.M., Kind, K.L., Kleemann, D.O., Plush, K.L., and Miller, D.R. (2014) A review of the genetic and epigenetic factors affecting lamb survival. *Animal Production Science* 54, 667–693.

Bunter, K.L., Johnston, D.J., Wolcott, M.L., and Fordyce, G. (2014) Factors associated with calf mortality in tropically adapted beef breeds managed in extensive Australian production systems. *Animal Production Science* 54, 25–36.

Canario, L., Cantoni, E., Le Bihan, E., Caritez, J.C., Billon, Y., Bidanel, J.P., and Foulley, J.L. (2006). Between-breed variability of stillbirth and its relationship with sow and piglet characteristics. *Journal of Animal Science* 84, 3185–3196.

Clutton-Brock, T.H., and Guinness, F.E. (1974) Behaviour of red deer (Cervus elaphus) at calving time. *Behaviour* 55, 287–300.

Curtis, S.E., 1970. Environmental-thermoregulatory interactions and neonatal piglet survival. *Journal of Animal Science* 31, 576–587.

Dargatz, D.A., Dewell, G.A., and Mortimer, R.G. (2004) Calving and calving management of beef cows and heifers on cow-calf operations in the United States. *Theriogenology* 61, 997–1007.

De Passillé, A.M.B., Rushen, J., and Harstock, T.G. (1988) Ontogeny of teat fidelity in pigs and its relation to competition at suckling. *Canadian Journal of Animal Science* 68, 325–338.

Deribe, G., Abebe, G., and Tegegne, A. (2014) Non-genetic factors influencing reproductive traits and pre-weaning mortality of lambs and kids under small-holder management, Southern Ethiopia. *Journal of Animal and Plant Sciences* 24, 413–417.

Dwyer, C.M. (2003) Behavioural development in the neonatal lamb: Effect of maternal and birth-related factors. *Theriogenology* 59, 1027–1050.

Dwyer, C.M. (2008a) The welfare of the neonatal lamb. *Small Ruminant Research* 76, 31–41.

Dwyer, C.M. (2008b) Genetic and physiological effects on maternal behavior and lamb survival. *Journal of Animal Science* 86, E246–258.

Dwyer, C.M., and Lawrence, A.B. (2005) A review of the behavioural and physiological adaptations of extensively managed breeds of sheep that favour lamb survival. *Applied Animal Behaviour Science* 92, 235–260.

Dwyer, C.M., Lawrence, A.B., Bishop, S.C., and Lewis, M. (2003) Ewe-lamb bonding behaviours at birth are affected by maternal undernutrition in pregnancy. *British Journal of Nutrition* 89, 123–136.

Edwards, S.A. (2002) Perinatal mortality in the pig: Environmental or physiological solutions? *Livestock Production Science* 78, 3–12.

Edwards, S.A. (2005) Product quality attributes associated with outdoor pig production. *Livestock Production Science* 94, 5–14.

Edwards, S.A., and Baxter, E.M. (2015) Piglet mortality: Causes and prevention. In: Farmer, C. (Ed.) *The gestating and lactating sow*. Wageningen Academic Publishers, Wageningen, Netherlands. pp. 253–269. DOI 10.3920/978-90-8686-803-2_11

Edwards, S.A., Smith, W.J., Fordyce, C., and Macmenemy, F. (1994) An analysis of the causes of piglet mortality in a breeding herd kept outdoors. *Veterinary Record* 135: 324–327.

Eriksson, S., Nasholm, A., Johansson, K., and Philipsson, J. (2004) Genetic parameters for calving difficulty, stillbirth, and birth weight for Hereford and Charolais at first and later parities. *Journal of Animal Science* 82, 375–383.

Fisher, M. (2003) New Zealand farmer narratives of the benefits of reduced human intervention during lambing in extensive farming systems. *Journal of Agricultural and Environmental Ethics* 16, 77–90.

Haughey, K.G. (1980) The effect of birth injury to the foetal nervous system on the survival and feeding behaviour of lambs. In: Wodzicka-Tomasczewska, M., Edey, T.N., and Lynch, J.J. (Eds.) *Reviews in rural science No. 4.* University of New England, Armidale. pp. 109–111.

Haughey, K.G. (1993) Perinatal lamb mortality – Its investigation, causes and control. *Irish Veterinary Journal* 46, 9–28.

Herpin, P., Damon, M., and Le Dividich, J. (2002) Development of thermoregulation and neonatal survival in pigs. *Livestock Production Science* 78, 25–45.

Hodgson, J.C. (1994). In: Gyles, C.L. (Ed.) *Escherichia coli in domestic animals and humans.* CAB International, Wallingford, Oxon, UK. pp. 135–150.

Hodgson, J.C., Barclay, G.R., Hay, L.A., Moon, G.M., and Poxton, I.R. (1995) Prophylactic use of human endotoxin-core hyperimmune gamma-globulin to prevent endotoxemia in colostrum-deprived, gnotobiotic lambs challenged orally with Escherichia coli. *FEMS Immunolology and Medical Microbiology* 11, 171–180.

Hoppe, S., Brandt, H.R., Erhardt, G., and Gauly, M. (2008) Maternal protective behaviour of German Angus and Simmental beef cattle after parturition and its relation to production traits. *Applied Animal Behaviour Science* 114, 297–306.

Jarvis, S., Lawrence, A.B., Mclean, K.A., Chirnside, J., Deans, L.A., and Calvert, S.K. (1998) The effect of environment on plasma cortisol and beta-endorphin in the parturient pig and the involvement of endogenous opioids. *Animal Reproduction Science* 52, 139–151.

Jensen, P. (1986) Observations on the maternal behaviour of free-ranging domestic pigs. *Applied Animal Behaviour Science* 16, 131–142.

Kamber, R., Farah, Z., Rusch, P., and Hassig, M. (2001) Studies on the supply of immunoglobulin G to newborn camel calves (Camelus dromedarius). *Journal of Dairy Research* 68, 1–7.

Kaufmannn, B.A. (2003) Differences in perception of causes of camel calf losses between pastoralists and scientists. *Experimental Agriculture* 39, 363–378.

Kaufmannn, B.A. (2005) Reproductive performance of camels (Camelus dromedarius) under pastoral management and its influence on herd development. *Livestock Production Science* 92, 17–29.

Kelly, R.W. (1992) Lamb mortality and growth to weaning in commercial Merino flocks in Western Australia. *Australian Journal of Agricultural Research* 43, 1399–1416.

Kilbride, A.L., Mendl, M., Statham, P., Held, S., Harris, M., Cooper, S., and Green, L.E. (2012) A cohort study of preweaning piglet mortality and farrowing accommodation on 112 commercial pig farms in England. *Preventive Veterinary Medicine* 104, 281–291.

Lawrence, A.B., Petherick, J.C., Mclean, K.A., Deans, L.A., Chirnside, J., Vaughan, A., Clutton, E., and Terlouw, E.M.C. (1994) The effect of environment on behavior, plasma-cortisol and prolactin in parturient sows. *Applied Animal Behaviour Science* 39: 313–330.

Lent, P.C. (1974) Mother-infant relationships in ungulates. In: Geist V., and Walther F. (Eds.) *The behaviour of ungulates and its relationship to management.* IUCN Publication No. 24, Morges, Switzerland. pp. 14–55.

Lidfors, L., and Jensen, P. (1998) Behaviour of free ranging beef cows and calves. *Applied Animal Behaviour Science* 20, 237–247.

Lidfors, L., Moran, D., Jung, J., Jensen, P., and Castren, H. (1994) Behaviour at calving and choice of calving place in cattle kept in different environments. *Applied Animal Behaviour Science* 42, 11–28.

Matheson, S.M., Bünger, L., and Dwyer, C.M. (2012) Genetic parameters for fitness and neonatal behavior traits in sheep. *Behavior Genetics* 42, 899–911.

McGlone, J.J., and Hicks, T.A. (2000) Farrowing hut design and sow genotype (Camborough-15 vs 25% Meishan) effects on outdoor sow and litter productivity. *Journal of Animal Science* 78, 2832–2835.

Mellado, M., Foote, R.H., and Detellitu, J.N. (1991) Effects of age and season on mortality of goats due to infections and malnutrition in north east Mexico. *Small Ruminant Research* 6, 159–166.

Mellado, M., Vera, T., Meza-Herrera, C., and Ruiz, F. (2000) A note on the effect of air temperature during gestation on birth weight and neonatal mortality of kids. *Journal of Agricultural Sciences* 135, 91–94.

Mellor, D.J., and Diesch, T.J. (2006) Onset of sentience: The potential for suffering in fetal and newborn farm animals. *Applied Animal Behaviour Science* 100, 48–57.

Mellor, D.J., and Gregory, N.G. (2003) Responsiveness, behavioural arousal and awareness in fetal and newborn lambs: Experimental, practical and therapeutic implications. *New Zealand Veterinary Journal* 51, 2–13.

Mellor, D.J., and Stafford, K.J. (2004) Animal welfare implications of neonatal mortality and morbidity in farm animals. *Veterinary Journal* 168, 118–133.

Moussouttas, M., Abubakr, A., Grewal, R.P., and Papamitsakis, N. 2006. Eclamptic subarachnoid haemorrhage without hypertension. *Journal of Clinical Neuroscience* 13, 474–476.

Mukasa-Mugerwa, E. (1981) The camel (Camelus dromedaries): A bibliographical review. *ILCA Monographs*, number 5. International Livestock Centre for Africa, Addis Ababa, Ethiopia.

Nix, J.M., Spitzer, J.C., Grimes, L.W., Burns, G.L., and Plyler, B.B. (1998) A retrospective analysis of factors contributing to calf mortality and dystocia in beef cattle. *Theriogenology* 49, 1515–1523.

Nowak, R. 1991. Development of mother discrimination by single and multiple newborn lambs. *Animal Behaviour* 42, 357–366.

Nowak, R., and Poindron, P. (2006) From birth to colostrum: Early steps leading to lamb survival. *Reproduction, Nutrition, Development* 46, 431–446.

Nowak, R., Poindron, P., Le Neindre, P., and Putu, I.G. (1987) Ability of 12-hour old merino and crossbred lambs to recognise their mothers. *Applied Animal Behaviour Science* 17, 263–271.

Nowak, R., Porter, R. H., Lévy, F., Orgeur, P., and Schaal, B. (2000) Role of mother-young interactions in the survival of offspring in domestic mammals. *Reviews of Reproduction* 5, 153–163.

Ogata, Y., Nakao, T., Takahashi, K., Abe, H., Misawa, T., Urushiyama, Y., and Sakai, J. (1999) Intrauterine growth retardation as a cause of perinatal mortality in Japanese black beef calves. *Journal of Veterinary Medicine, Series A: Physiology, Pathology, Clinical Medicine* 46, 327–334.

Ojha, S., Robinson, L., Yazdani, M., Symonds, M.E., and Budge, H (2013) Brown adipose tissue genes in pericardial adipose tissue of newborn sheep are downregulated by maternal nutrient restriction in late gestation. *Pediatric Research* 74, 246–251.

Patterson, D.J., Bellows, R.A., Burfening, P.J., and Carr, J.B. (1987) Occurrence of neonatal and postnatal mortality in range beef cattle. 1. Calf loss incidence from birth to weaning, backward and breech presentations and effects of calf loss on subsequent pregnancy rate of dams. *Theriogenology* 28, 557–571.

Poindron, P., Levy, F., and Nowak, R. (2007) Behaviour of the mother and the neonate in mammals: Physiological factors of activation. *Productions Animales* 20, 393–407.

Pollard, J.C. (2003) Research on calving environments for farmed red deer: A review. *Proceedings of the New Zealand Society of Animal Production* 63, 247–250.

Putu, I.G., Poindron, P., and Lindsay, D.R. (1988) Early separation of ewes from the birth site increases lamb separations and mortality. *Proceedings of the Australian Society for Animal Production* 17, 298–301.

Ramler, J.P., Hebblewhite, M., Kellenberg, D., and Sime, C. (2014) Crying wolf? A spatial analysis of wolf location and depredations on calf weight. *American Journal of Agricultural Economics* 96, 631–656.

Riart, G.R., Edwards, S.A., and English, P.R. (2000) Assessment of hypothermia in outdoor newborn piglets and comparison with an indoor system. In: *Proceedings of the British Society of Animal Science*. p.139.

Riley, D.G., Chase, C.C., Olson, T.A., Coleman, S.W., and Hammond, A.C. (2004). Genetic and nongenetic influences on vigor at birth and preweaning mortality of purebred and high percentage Brahman calves. *Journal of Animal Science* 82, 1581–1588.

Roehe, R., Shrestha, N.P., Mekkawy, W., Baxter, E.M., Knap, P.W., Smurthwaite, K.M., Jarvis, S., Lawrence, A.B., and Edwards, S.A. (2009) Genetic analyses of piglet survival and individual birth weight on first generation data of a selection experiment for piglet survival under outdoor conditions. *Livestock Science* 121, 173–181.

Roehe, R., Shrestha, N.P., Mekkawy, W., Baxter, E.M., Knap, P.W., Smurthwaite, K.M., Jarvis, S., Lawrence, A.B. and Edwards, S.A. (2010) Genetic parameters of piglet survival and birth weight from a two-generation crossbreeding experiment under outdoor conditions designed to disentangle direct and maternal effects. *Journal of Animal Science* 88, 1276–1285.

Rooke, J.A., Sinclair, A.G., Edwards, S.A., Cordoba, R., Pkiyach, S., Penny, P.C., Penny, P., Finch, A.M., and Horgan, G.W.

(2001) The effect of feeding salmon oil to sows throughout pregnancy on pre-weaning mortality of piglets. *Animal Science* 73, 489–500.

Sangild, P.T., Fowden, A.L., and Trahair, J.F. 2000. How does the foetal gastrointestinal tract develop in preparation for enteral nutrition after birth? *Livestock Production Science* (66) 141–150.

Sawalha, R.M., Conington, J., Brotherstone, S., and Villanueva, B. (2006) Analyses of lamb survival of Scottish Blackface sheep. *Animal* 1, 151–157.

Schwedt, T.J., Matharu, M.S., and Dodick, D.W. (2006) Thunderclap headache. *Lancet Neurology* 5, 621–631.

Singh, M.K., Dixit, A.K., Roy, A.K., and Singh, S.K. (2014) Analysis of prospects and problems of goat production in Bundelkhand region. *Range Management and Agroforestry* 35, 163–168.

Soares, C.M., Simoes, S.V.D., Medeiros, J.M.A., Riet-Correa, F., and Pereira, J.M. (2010) Passive immunity, colostrum ingestion, and mortality of Moxoto kids raised under extensive and intensive breeding systems. *Arquivo Brasileiro de Medicina Veterinaria e Zootecnia* 62, 544–548.

Stehulova, I., Spinka, M., Sarova, R., Machova, L., Knez, R., and Firla, P. (2013) Maternal behaviour in beef cows is individually consistent and sensitive to cow body condition, calf sex and weight. *Applied Animal Behaviour Science* 144, 89–97.

Stephenson, R.G.A., Suter, G.R., and Le Feuvre, A.S. (1984) Reduction in the effects of heat stress on lamb birth weight and survival by provision of shade. In: Lindsay, D.R., and Pearce, D.T. (Eds.) *Reproduction in Sheep*. Australian Academy of Science, Canberra. pp. 223–225.

Stevens, D., Alexander, G., and Lynch, J.J. (1982) Lamb mortality due to inadequate care of twins by merino ewes. *Applied Animal Ethology* 8, 243–252.

Turner, S.P., and Lawrence, A.B. (2007) Relationship between maternal defensive aggression, fear of handling and other maternal care traits in beef cows. *Livestock Science* 106, 182–188.

Turner, S.P., Jack, M.C., and Lawrence, A.B. (2013) Precalving temperament and maternal defensiveness are independent traits but precalving fear may impact calf growth. *Journal of Animal Science* 91, 4417–4425.

Van den Brand, H., van Enckewort, L.C.M., van der Hoeven, E.M., and Kemp, B. (2009) Effects of dextrose plus lactose in the sows diet on subsequent reproductive and within litter birth weight variation. *Reproduction in Domestic Animals* 44, 884–888.

Wallenbeck, A., Rydhmer, L., and Thodberg, K. (2008) Maternal behaviour and performance in first-parity outdoor sows. *Livestock Science* 116, 216–222.

Wass, J.A., Pollard, J.C., and Littlejohn, R.P. (2003) A comparison of the calving behaviour of farmed adult and yearling red deer (Cervus elaphus) hinds. *Applied Animal Behaviour Science* 80, 337–345.

chapter nine

Transport of livestock from extensive production systems

Clive Phillips

INTRODUCTION

Livestock are kept extensively because locally-available resources are not sufficiently concentrated or readily available for intensive production. In particular, natural feed resources are usually sparsely available and of poor quality, because the climate, terrain, and/or soil quality are unable to sustain prolific plant growth. In addition labour and capital resource inputs may be limited, relative to land area. Ruminant livestock predominate because of their ability to digest coarse fibre that offers insufficient nutrition for monogastric livestock, such as pigs and poultry. This they do by virtue of the large chamber or modified stomach in their gastrointestinal tract that contains micro-organisms to assist in the breakdown of the fibre, which is then digested to release the necessary nutrients for survival. As well as ruminants, in some parts of the world other animals, such as horses and elephants, thrive in extensive pastureland because of their ability to digest the fibre in an enlarged chamber that is part of their hindgut.

Just as the rangelands only support limited plant growth, so they also have not generally supported major human populations. As a result, transport of the animals, alive or dead, and/or their products is necessary for their potential to contribute to human nutrition to be realized. Also as humans congregate increasingly in cities away from the rangelands, it is inevitable that animals and their products must travel long distances to reach the consumers. This has been an evolving process over time, as transport improvements have enabled livestock to be kept in remote locations from the market that they serve, for example removal of dairy production from cities following the development of a train network to convey milk to the cities in Victorian England.

This chapter deals primarily with the transport of live animals, which is often

Centre for Animal Welfare and Ethics, School of Veterinary Science, University of Queensland, Australia preferred to transport of carcasses or products because of its low cost, but it creates some major welfare issues on which this book is focused. Live animals can travel unrefrigerated and unprocessed, getting onto and off the transport vehicle or vessel themselves, and often continuing to grow and gain weight during the journey. Inexpensive conversions of simple vessels, for example container or car-carrying ships, can be quite easily refurbished for this purpose. However, livestock welfare is likely to be compromised by use of vehicles and vessels that have not been specifically constructed for the purpose. This is particularly the case for animals that have had little contact with humans, such as those on rangelands.

Long-distance livestock transport mostly follows a route from the southern hemisphere, where there are extensive rangelands and labour is relatively cheap, to the northern hemisphere, where there are more concentrated centres of population. Typical of this is the sea transport of large numbers of sheep and cattle from Australia to Asia.

TRANSPORT METHODS

The type of transport method selected for livestock depends on how far the animals are being transported, the infrastructure available, and the ability of the animals to withstand the transport. Transport is price sensitive and the competition from other commodities for vehicles and vessels is a key factor in determining the transport method. In recent years there has been less use of rail travel for livestock transport, despite the fact that the welfare standards are often better than for truck transport. The main reason for

this is that livestock transported by rail usually have to be also conveyed by truck, to and from the train, increasing the handling requirements which are a major part of the cost.

Historically, transport by droving was one of the most common methods of moving livestock to markets. It is currently less used than in the past, being slow to deliver animals, labour consuming, and potentially stressful for the animals. Some countries, such as Australia, were well set up with long-distance droving routes, which are preserved to this day but rarely used for their original purpose. Droving of animals from rural districts to markets in the large cities in European countries has largely discontinued with the development of opportunities for more rapid conveyance by vehicle. Droving is still used in some developing countries, for example Africa, where cattle and sheep are sent from Sudan to Saudi Arabia via Eritrea. Some cattle are droved for long journeys in Brazil to finishing farms. Ensuring adequate fodder both en route and for overnight stops, when the animals are resting, is important for their welfare, as is water availability. Hoof problems and lameness often arise if the animals are droved for too long or on dry, stony tracks.

As well as droving, animals may be moved away from markets in cars, small trucks, or bicycles in developing countries. In vehicles they may be tied onto roofs, or held in the trunk or actually within a car. Separated from their social group and in close proximity to humans, stress levels will be high until well after the animal arrives at its destination.

Long distant transport by sea has been growing despite intense competition for container vessels worldwide. In part this has been due to recent improvements in international trading opportunities, with an emphasis on free trade fostered by bodies such as the World Trade Organisation. Livestock are often covered by bilateral agreements that aim to improve prices reached for the vendors and provide a supply of reasonably-priced meat for the consumer. Livestock transported by ship are downloaded from the truck in which they arrived at the port, and fed and watered if it is over a long distance, which is usually the case for extensively-reared livestock. There are about 40,000 merchant ships used worldwide for today's growing international commodity trade, however, only a small fraction of these are used for livestock (Langewiesche, 2010). The ability to regulate conditions for the animals is limited, with most of the steel cargo vessels flying under a flag of convenience and sailing in unregulated territories for most of their voyage. The crew are largely responsible for day-to-day care of the animals' welfare and are sourced from developing countries where there are few if any standards for animal welfare. They are principally operating to contain mortality to an economically-acceptable level. Refitting of the vessel is largely conducted en route to avoid berthing fees, which can cause disruption to sensitive livestock. Cleaning of the vessel and in particular the livestock pens is conducted on return routes after delivery of the livestock cargo.

On land, the dispersed nature of livestock farms in remote extensive regions favours transport by trucks, rather than rail, and this has been supported by the instalment of metalled roads to many rural areas over the course of the twentieth century. For example, northern Australia came to be serviced by a metalled road around the perimeter that could be used to transfer cattle to sea ports for live export. The success of this system led to closure of abattoirs in the north of the country in the 1990s. Prior to this livestock had to be droved to the ports or slaughtered locally. The Second World War and the military threat from the north facilitated improvement of this road network. Similarly, improvement of the southern Australian road from east to west allowed sheep to be transported from eastern states to Fremantle for the sea journey to the Middle East. In Europe, the presence of markets on the town and city boundaries brought animals within full view of the public once conurbation surrounded the market, but increasingly these have been moved out of town, changed to online marketing, or direct sales to abattoirs.

As roads have improved, the extensive livestock-rearing regions of the world have increasingly been serviced by large transport vehicles, sometimes capable of moving up to 200 head of cattle in three interlinked trailers with two decks each. These "road trains" are efficient means of conveying livestock but on unmetalled roads they can create dust and trailer movement that potentially stress the animals. Animals conveyed by this method have a high prevalence of respiratory and other infections after arrival, in particular bovine respiratory disease. Increased risk occurs with low temperatures, long distances travelled, poor ventilation, overstocking, mixing with other groups, and, for male calves, castration (Cernicchiaro et al., 2012). Overstocking, as well as increasing heat stress, may lead to ventilation ports becoming blocked, contributing to respiratory disease.

Air travel is about four times as expensive per kilometre travelled as ship transport but may still be economic if animals are high value, for example calves with high meat or milk production potential. This does not usually apply to rangeland livestock. Therefore only a very small proportion of cattle and sheep are transported by air, for example 1% of cattle and 2% of sheep in Australia (MLA, 2013). However, a much larger proportion of goats is sent from Australia to Asia, mostly to Malaysia, 98%. These are usually feral goats rounded up from northern states and therefore prone to transport stress. The short duration of air travel suits them better than road or rail travel. These animals are particularly stress prone because of their infrequent contact with humans. The high mortality rate on ships used to be tolerated because the animals are seen as a pest species in the Australian outback. However, airplanes are now used more frequently with benefit to the animals' welfare.

LONG-DISTANCE TRANSPORT OF LIVESTOCK

The largest trade in the world of transport of livestock by sea is that of sheep coming into Asia. There is a particularly high demand for sheep and goat meat in the Middle East at the time of the Haj festival. About six million animals in total are slaughtered, with about 42% coming from Australia, 43% from the Horn of Africa, including many animals from the Sudan, and about 15% from Eastern Europe (Abbas et al., 2014). Thus the majority of animals have come from distant grazing lands and change in conditions and length of the journey, several weeks, make the animals prone to stress and disease. Outbreaks of Rift Valley fever have occurred in animals from Africa, which have occasionally halted the trade, and sheep from Australia have been detected with pustular dermatitis, which has also on occasion led to the trade being halted.

Since its development in the 1990s, the long-distance transport of sheep and cattle between Australia and Asia has been punctuated by regular concerns about the health of the animals on the part of importing nations and concerns about the animals' welfare by the exporting nations. For example, trade between Australia and Saudi Arabia was suspended in 1989, 1991, and 2003, following concerns by the Australian government about the welfare of sheep on board (1991) and by the Saudi government about disease levels in the sheep (1989/1990 and 2003; Phillips, 2015). In 2005 the Australian government signed a Memorandum of Understanding with the Saudi authorities providing a quarantine facility near Jeddah to be used if required, which allowed the re-opening of the trade after the 2003 suspension. High temperatures and humidity in the Persian Gulf region have also been a regular problem, mainly for shipments leaving Australia in the southern hemisphere winter and entering the Gulf in the northern hemisphere summer, with cattle in particular unable to cope with the sudden increase in temperature. Mortalities of cattle have occasionally reached 44% (MV Becrux in 2002), and those of sheep, 10% (Cormo Express in 2003) (Phillips, 2015, p. 101).

Sheep arriving either on foot or in trucks to Eritrean, Djibouti or Somalian ports for ship transport across to the

Arabian Peninsula are often in poor condition. Those that have been droved are in best condition, those on the trucks are often overcrowded, injured, and/or heat stressed (Abbas et al., 2014). Pneumonia is a major cause of mortality and a survey of 1,400 livestock of different types in Djibouti found that 8% of consignments were rejected, often camels with pox. Camels are difficult to transport and have to be tied down on board. A total of 1.4 million livestock leave this port for the Arabian Peninsula each year. Slightly more sheep, two million, leave the Australian shores annually, most bound for the Middle East, although this number has declined recently following prolonged drought in Australia.

As outlined above, extensive rangelands are usually distant from their markets, and in the view of stakeholders in the industry, long distant transport represents one of the most serious welfare concerns for these animals (Pines et al, 2007). For cattle, the process starts with mustering by stockpeople on horses, motorbikes, in vehicles, or planes and helicopters. In the most extensive rangelands an airplane first flushes them out of wooded country, then a helicopter may usher them along fencelines or traditional paths used by cattle towards the holding pens, followed by stockpeople on horseback, motorbike, or vehicle on the ground to take them to the pens, working at the front and back of the mob. If they arrive hot and flustered, they should be held outside of the pens and allowed to enter slowly of their own accord. Vehicles should never be used to hit animals that attempt to evade the muster. Rushing the animals can lead to fatalities from heat stress. Once in the pens they are usually held overnight to

calm down and recover before being taken in a vehicle to an assembly depot, where they wait for a few days until all the consignment of animals has arrived. Stockpeople should be trained to show mindful and gentle handling of animals at all times, minimizing stress to the animals and themselves. They should never engage in unacceptable handling practices that may cause distress, pain, or injury, such as excessive use of an electronic prodder or goad. Rough handling causes some animals to baulk, resulting in a slower and more stressful procedure for the animal. Ideally they should let animals walk at their own pace (Animal Welfare Standards, 2015).

Many transport and marketing companies like to prevent animals feeding before they are trucked to minimize accumulation of excreta in the truck. In Australia the maximum permitted time off water is about 48 hours, depending on the type of animal. However, this is likely to have penalties for rumen microbial activity, with a potential increase in enteropathogenic bacteria (Hogan et al., 2007). The tiredness of cattle transported long distances is evident on arrival at their destination if they go straight to lie down without feeding and drinking. The quality of the journey is an important factor here. The truck driver should be careful to minimize sharp cornering and fierce braking, especially on unmetalled roads that create the dust that particularly affects the rear carriages of long road trains. The truck used should have a non-slip floor with some floor litter or a means of removing faeces. Stocking density is often too high, which predisposes livestock to heat stress. The stocking density is determined from the animals' weight in established standards, taking into account fleece length in sheep, and whether animals are pregnant, or horned. The length of the journey must also be considered, with stocking density relaxed for longer journeys.

The assembly depot is a particularly important part of the process for sheep, who need to get used to eating pellets and the high stocking densities employed on ships. During this time animals are mixed from a number of farms, but it is important to keep entire males separate to avoid fighting. If the journey is particularly long it may be necessary to offload the animals to allow them to eat and drink, however, the stress of doing this has to be weighed against the benefit of allowing them to replenish nutrients. Loading and unloading will be less stressful to animals if they are allowed to walk on and off calmly, with ramps of no greater than 15–20° and use of batons on the ramps to allow the animals a good purchase if it is slippery. The use of a neck hoist and crane to winch Australian cattle onto a ship when being transported between Indonesian islands causes damage to some cattle, in the form of heart failure and failure to raise their head normally afterwards (WSPA, 2013).

Assembly depots are usually near the ports, and another journey takes them to the loading bays for entry into the ship. Loading and offloading 50–100,000 animals in one day requires good coordination, particularly the loading as the anticipated departure time may not be known until late in the procedure because of changing weather and tides. Availability of sufficient trucks to transport the animals quickly on and off the vessel often cannot be guaranteed, leading to livestock waiting on or off ship for long periods. During loading, regular inspection for sick or injured animals or those that are too thin (or less commonly too fat) to undertake the journey is important, and this is undertaken in Australia by veterinary officials of the Australian Quarantine and Inspection Service. Animals may be encouraged to move more rapidly by use of electronic prodders, which stresses them and means that they take longer to settle on board. This should only be undertaken in an emergency, if the stress to other waiting animals exceeds the stress to the individual animal that is to be electrocuted.

After arrival on board, cattle and sheep are allowed to settle for a few hours before food and water is provided. Failure to eat is a major problem for sheep, particularly those that are overweight as they are able to lose weight initially to survive and don't have the same appetite as thinner sheep.

Cattle and sheep commonly spend one to two weeks on board ship, but this can be extended to three to four weeks for distant destinations, after which they are offloaded and may be taken to a feedlot to gain further weight or direct to an abattoir for slaughter. The major trade in cattle from northern Australia to Indonesia is mainly in young cattle, which are finished in feedlots on a grain-based diet over a period of 10–50 days. Short periods in the feedlot are of little value as it takes ruminant livestock approximately two weeks to adapt to their new conditions and diet before they start to gain weight.

STRESS TO LIVESTOCK DURING TRAVEL

Even short distance ship journeys, such as between the British Isles and the

European continent are a cause of stress to livestock, as evidenced by immuno-suppression, weight loss, and physiological measures indicating bruising (Earley and Murray, 2010; Earley et al., 2011, 2012). Longer travel potentially results in even greater stress to livestock and it is important to have measures in place to recognize poor performance. However, the large numbers transported in some systems, for example up to 80,000 sheep in ships from Australia to the Middle East, make detailed measurements difficult. Lack of access to observe animals may also present problems for livestock in trucks. The simplest and most commonly used measure for truck and sea journeys is mortality rate, but this is obviously unable to detect stresses that do not usually kill the animals, for example ammonia. Mortality rates, however, have been measured for the Australian live export trade for many years and provide a measure that can be compared across time. These have declined for sheep over the last 10–15 years but remained relatively constant for cattle. Average levels are 0.14%, 1.0%, and 1.4% for cattle, sheep, and goats, respectively; least for cattle in part because the journeys are shorter, but also because they are less prone to inappetence and salmonellosis than sheep and goats (Richards et al., 1989; AHAW, 2011). There is some evidence that ship journeys are less stressful than journeys by truck (Hall et al., 1999), but much depends on the duration and quality of each. Undernutrition remains a regular problem for livestock, particularly sheep, during long-distance travel. This is often exacerbated by ammonia (Phillips et al., 2012; Pines and Phillips, 2013) and high stocking densities that limit movement around the pen, and it is commonly

accompanied by salmonellosis in sheep. Sheep travelling from the south to central Chile by ship often arrive in an under-nourished state (Tadich et al., 2009). Ship travel is used for these animals because of inadequate roads in the far south. Similarly, in journeys of about one week between the Indonesian islands or from South Africa to Mauritius, inadequate food and water are persistent problems (Menczer, 2008).

As well as feed and water deficiencies, livestock on board a ship or a truck are subjected to many conditions that can cause stress. Movement of the vessel or vehicle is a fundamental cause of stress. Livestock bumping into each other, or the vessel or vehicle, causes bruising, which downgrades a carcass and signifies that significant and painful damage has been done to the soft tissue. Livestock in a vehicle are primarily subjected to forward/backward and sideways (sway) movements, whereas those in a ship have up/down (heave) movements to contend with. Heave is the major movement causing motion sickness in humans and may cause malaise in sheep also (Santurtun et al., 2013, 2015). The rolling of a ship, which may be by up to 20° from a central pivot point, causes splaying of the feet and stepping movements as sheep attempt to maintain their balance, eventually tiring animals (Jones et al., 2010). As well as the rolling and heave movements of a ship, freak waves can cause slamming, an unpredictable shunt movement of the vessel as a wave hits directly broadside. Any extreme movements of a vessel or vehicle can cause sheep to bunch together, predisposing them to suffocation. The variability in forces on a livestock in a ship make it unlikely that they will adopt any particular direction of travel, but on trucks cattle at least have been recorded standing perpendicular to the direction of travel, perhaps to protect their head in the event of sudden movement (Tarrant et al., 1992). In both trucks and ships, driving quality can affect the animals' experiences, with sudden turns and rough roads being the major concerns in relation to trucks (Cockram et al., 2004), and the ability of the captain to avoid high seas the major concern in relation to ships. The ship's captain can also activate stabilizers to reduce a ship's rolling action, but this reduces speed and fuel efficiency. In contrast to the forces livestock experience in road and sea transport, rail transport creates largely surge forces, with little roll or heave, therefore providing less impact on the animals than trucks or ships.

Heat stress is a problem for cattle, especially large, entire males, and sheep, with mature rams being most susceptible. Cattle from the south of Australia in winter have a long coat and when they enter summer in the northern hemisphere they are especially at risk of heat stress. Persistent wet bulb temperatures of 30°C (Pines and Phillips, 2011) to 34°C (Beatty et al., 2006, 2007) are relatively common in ships undertaking the journey from Australia to the Middle East, with no opportunity for the animals to recover in lower temperatures at night. In the worst heat stress disasters of livestock transported long distances by ship there were mortalities of 15,000 and 30,000 Australian sheep and 10,000 New Zealand sheep (Phillips, 2015). Despite the fact that there have been some major sheep disasters, cattle are more susceptible to heat stress than sheep by virtue of their larger body mass (Caulfield et al.,

2013). Wetting cattle with a hose using sea water can temporarily alleviate heat stress in cattle but it will also increase humidity, potentially causing heat stress later. High stocking densities increase the heat load that animals must endure and reduce their opportunities to find a well-ventilated part of the ship. One effect of heat stress is to increase water intake to make up for increased water losses when the animals are panting (cattle and sheep) or sweating (cattle only) (Salama et al., 2016, Chapter 2; Revell, 2016, Chapter 3). This increases urination, which dampens the bedding, if any is used, further increasing humidity. At the same time feed intake declines, and with it the heat generated in ruminant livestock by digestion of the feed. There is evidence of respiratory alkalosis and renal dysfunction in sheep experiencing temperatures typical of those to which sheep are subjected when travelling from Australia to the Middle East (Stockman et al., 2011). Accurate monitoring of temperatures on a number of decks is important in reducing the risk of mortality as a result of heat stress. A particular high-risk time is when the ship is in port, as wind speed is usually less than out at sea. This emphasizes the importance of rapid and effective loading and offloading.

Adequate ventilation is essential for any livestock transporter. Ships may have natural ventilation on upper decks and forced ventilation below sea level, with 20–30 air changes per hour and a wind speed of 0.5 m/s needed to avoid excessive ammonia accumulation (Pines and Phillips, 2011). Ventilation systems need to be carefully designed, and excessive wind speed at the point of arrival of the air at sheep decks can cause pneumonia in nearby sheep. This has been a major cause of mortality in sheep, for example those travelling from New Zealand (Black et al., 1994). High stocking densities limit movement of sheep around the pens (Pines and Phillips, 2013), preventing animals controlling their own micro-climate.

Ammonia accumulation is a major problem on ships and can accumulate within a few days (Pines et al., 2007). Levels of 59 ppm (Pines and Phillips, 2011) and 187 ppm (Earley et al., 2011) have been recorded, with the commonly accepted "no-go" limit for humans being 50 ppm. The ammonia comes from urea in faeces and urine, which converts to ammonium hydroxide when in contact with a moist surface, especially the lungs, eyes, and mouth of animals. This irritates the mucosal surface and causes infections in the lungs and eyes. The animals cry, sneeze, and cough (Pines and Phillips, 2013), and the eye inflammation causes a progressive conjunctivitis that renders many unable to see out of one or both eyes by the end of the journey (Pines and Phillips, 2013). Ammonia could be controlled by limiting protein content of the feed or by adding zeolites to it, or the bedding, to reduce volatilization of nitrogenous compounds from urine and faeces.

Both heat stress and ammonia accumulation are primarily a function of stocking density and temperature/humidity in the pen. Stocking density should be at least enough for an animal to lie down, turn around, and move to eat (Petherick and Phillips, 2009; 0.33 m^2 for sheep and 1.25 m^2 for 300 kg cattle). The stocking densities used are usually greater for livestock on ships than trucks, but less than for those in feedlots or sale yards. With such limited space there is a risk that animals that lie down may experience other animals closing over them so that they are unable to get up. Hence animals may be reluctant to lie down, for example in the early stages of a sea voyage. The high risk of mortality in sheep and cattle exported from Australia has led the industry to develop a method of controlling stocking density to limit heat stress risk (Caulfield et al., 2013). The provision of suitable bedding, at least for cattle who produce faeces with more moisture than sheep, will help to alleviate some of the effects of high stocking densities. However, care should be taken on decks with a risk of incursion of seawater. Wet bedding will seriously reduce welfare conditions for sheep. Hospital pens should be provided on board ships, in which a relaxed stocking density can assist a small number of animals in their recovery.

TRANSPORT STANDARDS

The growing scale of international trade in livestock means that standards, preferably international, will be of increasing importance to monitor and improve animal welfare to satisfy the growing demands of the general public. The public have been alerted to welfare issues surrounding the live export trade by activist exposés (Tiplady et al., 2012).

The World Organisation for Animal Health (OIE) has prepared standards for the welfare of transported livestock, which were originally guidelines (OIE, 2012). By the standards of many national governments, these are relatively lax, essentially because they have to gain the support of all OIE-supporting countries.

The OIE does not enforce the standards or monitor compliance with the standard, but leaves this to governments. It does, however, support training in the standards in developing countries (Animal Welfare Standards, 2015).

A major problem with international transport of livestock is that when in international waters the animals are not subject to any animal welfare legislation and may not even be traceable when they are in the importer's country. Australia has recently developed an Exporter Supply Chain Assurance Scheme (ESCAS), which has the following objectives:

1. Animals will be handled and processed at the internationally accepted standards for animal welfare established by the World Organisation for Animal Health (OIE) or better;
2. Exporters have control of the movement of animals within their supply chain;
3. Exporters can trace or account for animals through the supply chain; and
4. Exporters conduct independent verification and performance audits of their supply chains against these new requirements

In addition, Australia has a Live Export Accreditation Programme, which was introduced in 1997 and Australian Standards for the Export of Livestock (ASEL, 2011), which are occasionally reviewed and revised when industry research indicates that change is warranted.

The Food and Agriculture Organisation has produced Guidelines for Humane Handling, Transport, and Slaughter of Livestock, which have no legal status but provide advice on pursuing best practice.

The strictest standards for transport internationally are those of the European Union (Council Regulation, 2005). These cover people's responsibilities (including a certificate of competency for drivers), the types of transporters that can be used, journey times for different classes of stock, inspection protocols, fitness of animals for transport, and rest periods. They are difficult to enforce but because of this a mandatory on-board satellite navigation system is being developed. Guidance on import and transit rules for live animals and animal products from third countries that are within the OIE are also available (European Commission, 2010). In contrast, developing countries often have no standards or very limited standards, or standards that are hard or impossible to enforce.

In addition to these national and international standards, there are private standards that are available for use by retailers, for example within the Global GAP framework established in 1997.

FUTURE RESEARCH

The growing long-distance trade in livestock, especially to meet the needs of the demand for animal products in Asia, must be accompanied by adequate research if the serious concerns of the general public (e.g., Tiplady et al., 2012) are ever to be assuaged. The causes of, and solutions for, stress to the animals involved in long-distance transport need to be much better understood. On ships these include high ammonia concentrations, stocking densities, temperatures and humidity, the need for constant stepping during ship motion, lack of familiarity with novel feeds, people and other sheep, strange noise and lighting, mishandling, and so on. Combinations of stressors need to be tested, for example, how do high temperatures affect ammonia concentrations?

To date, most of the limited amount of research on long-distance shipping has been conducted in Australia, which is believed to have some of the best ships. Potentially harmful practices, such as the hoisting of cattle into ships in Indonesia by ropes around their necks, need evaluating. Detailed epidemiological surveys are required to identify major risk points, from the point of embarkation until the eventual slaughter of livestock. If necessary, this should include the management of dairy cows following importation from distant lands, for example from New Zealand to China. Stress during slaughter processes in importing countries, in particular the Muslim states that prevent stunning of cattle before slaughter, should be documented and targets devised for improvement. These could accompany OIE standards, the only operational standards in most developing countries.

The attitudes of people in both exporting and importing countries to the practices involved in long-distance transport to or from their nation should be evaluated. Pressure for change will only be possible with good scientific information and evidence that there is little public support for practices that are harmful to animals. Exporting and importing countries should ask themselves why, over 50 years after Ghandi is reputed to have said "The greatness of a nation and its moral progress can be judged by the way its animals are treated" (Ghandi, 1959), we have little or no tangible evidence that

standards have improved. The increase in long-distance transport in fact makes us suspect that standards have actually declined for livestock.

CONCLUSIONS

Transport is a major welfare concern for extensively-reared livestock because the distances are large and duration of travel long and often arduous. The various means of moving livestock all have welfare concerns and some of the longest journeys have had regular disasters with major mortality events. It is important for the reputation of the industry that more is done to improve the animal's welfare on board ship, in trucks, and during droving. This requires research, particularly for ship transport and droving, since little is known about these compared to our understanding of the impact of trucking livestock on their welfare.

REFERENCES

Abbas, B., Yousif, M.A., and Nur, H.M. (2014) Animal health constraints to livestock exports from the Horn of Africa. *Revue Scientifique et Technique (International Office of Epizootics)* 33, 711–721.

Animal Welfare Standards (2015) International Animal Welfare Project. Available at: www.Animalwelfarestandards.org (accessed 15 March, 2015).

Australian Standards for the Export of Livestock (ASEL) (2011) Version 2.3, April, 2011. Commonwealth of Australia, Australian Government, Department of Agriculture, Fisheries and Forestry, Canberra.

Beatty, D.T., Barnes, A., Taplin, R., McCarthy, M., and Maloney, S.K. (2007) Electrolyte supplementation of live export cattle to the Middle East. *Australian Journal of Experimental Agriculture* 47, (1)19–124.

Beatty, D.T., Barnes, A., Taylor, E., Pethick, D., McCarthy, M., and Maloney, S.K. (2006) Physiological responses of Bos taurus and Bos indicus cattle to prolonged, continuous heat and humidity. *Journal of Animal Science* 84, 972–985.

Black, H., Matthews, L.R., and Bremner K.J. (1994) The behaviour of male lambs transported by sea from New Zealand to Saudi Arabia. *New Zealand Veterinary Journal* 42, 16–23.

Caulfield, M.P., Cambridge, H., Foster, S.F., and McGreevy, P.D. (2013) Heat stress: A major contributor to poor animal welfare associated with long-haul live export voyages. *The Veterinary Journal* 199, 223–228.

Cernicchiaro, N., White, B.J., Renter, D.G., Babcock, A.H., Kelly, L., and Slattery, R. (2012) Associations between the distance traveled from sale barns to commercial feedlots in the United States and overall performance, risk of respiratory disease, and cumulative mortality in feeder cattle during 1997 to 2009. *Journal of Animal Science* 90, 1929–1939.

Cockram, M.S., Baxter, E.M., Smith, L.A., Bell, S., Howard, C.M., Prescott, R.J., and Mitchell, M.A. (2004) Effect of driver behaviour, driving events and road type on the stability and resting behaviour of sheep in transit. *Animal Science* 79, 165–176.

Council Regulation (EC) No 1/(2005) On the protection of animals during transport and related operations and amending Directives 64/432/EEC and 93/119/EC and Regulation (EC) No 1255/97. *Official Journal of the European Union* L 3/44. Available at: http://eurlex.europa.eu/LexUriServ/site/en/oj/(2005/l_003/l_003(20050105en00010044.pdf (accessed 21 January 2014).

Earley, B., and Murray, M. (2010) The effect of road and sea transport on inflammatory, adrenocortical, metabolic and behavioural responses of weanling heifers. *BioMedical Central Veterinary Research* 6, 1–13.

Earley, B., McDonnell, B., Murray, M., Prendiville, D.J., and Crowe, M.A. (2011) The effect of sea transport from Ireland to the Lebanon on inflammatory, adrenocortical, metabolic and behavioural responses of bulls. *Research in Veterinary Science* 91, 454–464.

Earley, B., Murray, M., Prendiville, D.J., Pintado, B., Borque, C., and Canali, E. (2012) The effect of transport by road and sea on physiology, immunity and behaviour of beef cattle. *Research in Veterinary Science* 92, 531–541.

EFSA Panel on Animal Health and Welfare (AHAW). (2011) Scientific opinion concerning the welfare of animals during transport. *European Food Safety Authority Journal* 9, 1–125.

European Commission (2010) *General Guidance on EU Import and Transit Rules for Live Animals and Animal Products from Third Countries*. European Commission, Health and Consumers Directorate-General, Brussels. http://ec.europa.eu/food/international/trade/guide_thirdcountries(2009_en.pdf (accessed 21 January 2014).

Ghandi, M. (1959) Quote, origin unknown but reputed to have been taken by Ghandi from Ramachandra Krishna Prabhu, *The Moral Basis of Vegetarianism* (1959). http://en.wikiquote.org/wiki/Mahatma_Gandhi

Hall, S.J.G., Broom, D.M., Goode, J.A., Lloyd, D.M., Parrott, R.F., and Rodway, R.G. (1999) Physiological responses of sheep during long road journeys involving ferry crossings. *Animal Science* 69, 19–27.

Hogan, J.P., Petherick, J.C., and Phillips, C.J.C. (2007) The nutritional impact on sheep and cattle of feed and water deprivation prior to and during transport. *Nutrition Research Reviews* 20, 17–28.

Jones, T.A., Waitt, C., and Dawkins, M.S. (2010) Sheep lose balance, slip and fall less when loosely packed in transit where they stand close to but not touching their neighbours. *Applied Animal Behaviour Science* 123, 16–23.

Langewiesche, W. (2010) The outlaw sea. In: Bradley, J. (Ed.) *The Penguin Book of the Ocean*. Penguin Group, Camberwell, Australia. pp 415–443.

Meat and Livestock, Australia (MLA). (2013) *Australian Livestock Export Industry Statistical Review*. 12 pp. Meat and Livestock, Australia, Sydney.

Menczer, K. (2008) Africa. In: Appleby, M. C., Cussen, V.A., Garcés, L., Lambert, L. A., and Turner, J. (Eds.) *Long distance transport and welfare of farm animals*. CABI, Wallingford, UK. pp. 182–211.

OIE (2016) OIE's achievements in animal welfare. http://www.oie.int/animal-welfare/animal-welfare-key-themes/ Accessed 22 February, 2016.

Petherick, J.C., and Phillips, C.J.C. (2009) Space allowances for confined livestock and their determination from allometric principles. *Applied Animal Behaviour Science* 117, 1–12.

Phillips, C.J.C. (2015) The trade in live cattle and sheep. In: *The animal trade*. CABI, Wallingford. In press.

Phillips, C.J.C., Pines, M.K., Latter, M., Muller, T., Petherick, J.C., Norman, S.T., and Gaughan, J.B. (2012) The physiological and behavioral responses of sheep to gaseous ammonia. *Journal of Animal Science* 90, 1562–1569

Pines, M., and Phillips, C.J.C. (2011) Accumulation of ammonia and other potentially noxious gases on live export shipments from Australia to the Middle East. *Journal of Environmental Monitoring* 13, 2798–2807.

Pines, M., Petherick, J.C., Gaughan, J.B., and Phillips, C.J.C. (2007) Stakeholders' assessment of welfare indicators for sheep and cattle exported by sea from Australia. *Animal Welfare* 16, 489–498.

Pines, M.K., and Phillips, C.J.C. (2013) Microclimatic conditions and their effects on sheep behavior during a live export shipment from Australia to the Middle East. *Journal of Animal Science* 91, 4406–4416.

Richards, R.B., Norris, R.T., Dunlop, R.H., and McQuade, N.C. (1989) Causes of death in sheep exported live by sea. *Australian Veterinary Journal* 66, 33–38.

Santurtun, E., Moreau, M., Marchant-Forde, J.N., and Phillips, C.J.C. (2015) Physiological and behavioral responses of sheep to simulated sea transport motions. *Journal of Animal Science*. In press.

Santurtun, E., Moreau, V., and Phillips, C.J.C. (2013) Behavioural responses of sheep to simulated sea transport motion. In: *Proceedings of the 47th Congress of the International Society for Applied Ethology*, Brazil. pp. 164.

Stockman, C.A., Barnes, A.L., Maloney, S.K., Taylor, E., McCarthy, M., and Pethick, D., (2011) Effect of prolonged exposure to continuous heat and humidity similar to long haul live export voyages in Merino wethers. *Animal Production Science* 51, 135–143.

Tadich, N., Gallo, C., Brito, M.L., and Broom, D.M. (2009) Effects of weaning and 48 h transport by road and ferry on some blood indicators of welfare in lambs. *Livestock Science* 121, 132–136.

Tarrant, P.V., Kenny, F.J., Harrington, D., and Murphy, M. (1992) Long distance transportation of steers to slaughter: Effect of spatial allowance on physiology, behavior and carcass quality. *Livestock Production Science* 30, 223–238.

Tiplady, C., Walsh, D.B., and Phillips, C.J.C. (2012) Cruelty to Australian cattle in Indonesian abattoirs - how the public responded to media coverage. *Journal of Agricultural and Environmental Ethics* 26, 869–885.

World Society for Protection of Animals (WSPA) (2013) Report to Indonesian Veterinary Medical Association/ Government of Indonesia: *Inter-island livestock transport* – April 2013. 20 pp. WSPA, London.

Index

animal trypanosomiasis (Nagana) 144–5
animals
  behaviour-based management practice 4
  distribution 4
  environmental conditions 12–15
  maintaining in zone of thermal wellbeing 46
  productivity 20–2
  salt intake during pregnancy 47–8
  treating endo-/ectoparasites 66
  visual monitoring 29
  see also named animals
anthrax 139
ASEL see Australian Standards for the Export of
    Livestock
Australian Standards for the Export of Livestock
    (ASEL) 197

biosecurity 136, 149
birth defects 86–7
bison 134, 146, 147
Bluetongue 151
bovine tuberculosis 139, 147–8
Brucellosis 146–7
buffalo 14

camels 163, 178–9
cattle
  breeding 67–8
  diseases 143, 146–8
  effect of predation on 110–12, 114–16,
    117–18
  environmental adaptation 3, 60–2
  feed management 24–6
  foraging ability 59
  low-stress herding 58
  neonatal mortality 163, 168–72
  number of head of cattle in US 109–10
  predation 161–2

shelter/cover 56, 67
stocking level 57
thermal stress 11–12, 13–28, 195
transportation of 189, 191, 192, 193, 195
water balance/availability 37, 38, 40, 40–2, 44, 47,
    53–4
climate change 3, 5, 11, 61, 88–9, 135, 151
cold chain 143, 146
cold stress (CS) 11–12, 13, 16, 17, 23–4, 26–7, 28
CS see cold stress

deer 163
discomfort 2
  grazing/management of extensive systems
    66–7
  under thermal stress 17–18
disease see health; pain, injury, disease
distress see fear/distress

East Coast Fever 143–4
ECTAD see Emergency Centre for Transboundary
    Animal Diseases
EID see electronic identification
electronic identification (EID) 152
elk 146, 147
Emergency Centre for Transboundary Animal
    Diseases (ECTAD) 149
Endangered Species Act 107
environment 1–2, 4, 8
  differences between extensive/intensive
    systems 53–6
  interaction with animal characteristics 12–15
  modifications to reduce thermal stress 26–7
  natural disasters 138, 139
epigenetics see genetics/epigenetics
ergot alkaloids 84–5
ESCAS see Exporter Supply Chain Assurance
    Scheme

Exporter Supply Chain Assurance Scheme (ESCAS)
197
extensive systems description 133–5

FAO see Food and Agriculture Organization
Farm Animal Welfare Council (UK) 2
fear/distress 2
    due to predation 119–23
    grazing/management 68
    under thermal stress 20
five freedoms see discomfort; fear/distress; normal
        behavior; pain, injury, disease; thirst, hunger,
        malnutrition
Food and Agriculture Organization (FAO) 149,
        197
food supply 3, 4–5
    challenges in extensive systems 54–6, 138
    feed schedule/supplements 24–6, 30
    feed/forage intake 40–1
    grazing/animal distribution 54–6, 63–6
    quality/quantity 55, 57
    site selection 55–6
foot-and-mouth disease 142
fungal endophytes 84

genetics/epigenetics
    coping with environmental toxins 87–8
    gene expression regulation under thermal stress
        22–4
    genetic markers/molecular breeding values 62,
        181
    improving thermal tolerance 3–4, 27–8
    maintaining water balance 46–8
    selecting for environment 46–7
GHG see greenhouse gas
Global GAP framework 197
Global Rinderpest Eradication Programme (GREP)
        146
goats
    effect of toxic plants 82, 87–8
    foraging ability 59
    neonatal mortality 163, 177–8
    stocking rates 57
    thermal stress 11–12, 14–15, 18, 19, 21–3
grazing 4, 53
    affect on five freedoms 62–8
    challenges in extensive systems 53–6
    concerns/recommendations 69–70
    continuing research on 68–9
    distribution 57–8
    foraging challenges 54–6
    kind/class of animal 59–62
    low-stress stockmanship 69
    management 48–9, 56–62
    season of use 58
    shelter/cover 56

stocking rate 4, 56–7, 142
    water as critical 53–4
Greater Yellowstone Area (GYA) 146
greenhouse gas (GHG) 135
GREP see Global Rinderpest Eradication
        Programme
GYA see Greater Yellowstone Area

health 132–3
    animal-side tests 152
    biosecurity concerns 136
    challenges 6–7
    control of 140, 148–9
    description of extensive livestock system 133–5
    disease-resistant stock 136
    exposure to diseases 136–7
    factors influencing disease risk 137–43
    future knowledge 150–3
    grazing management 136
    hazards affecting 137–41
    hoof problems 189
    identification and action 135
    impact on food security/safety 153
    logistical problems 135
    mitigating factors for disease risk 141–3
    One Health paradigm 149–50
    provision of supplementary vitamins/minerals
        138
    skills of stockpersons 135–6, 140–1, 142–3, 153
    species under consideration 134
    specific diseases 143–8
    surveillance of animals 136
    use of antimicrobial/antiparasitic therapy 136
    use of biosecurity/biocontainment 149
    use of digital/smart technology 152
    vaccination 136, 139, 140, 146, 147, 149
    worldwide areas 134–5
    see also pain, injury, disease
heat stress (HS) 11, 12, 13, 16, 17–19, 21–2, 23–4,
        25–6, 138, 195
horses 81, 82
HS see heat stress
hunger see thirst, hunger, malnutrition

IBMP see Interagency Bison Management Plan
injury see pain, injury, disease
Interagency Bison Management Plan (IBMP) 147
Intergovernmental Panel on Climate Change 11

larkspurs (Delphinium species) 85–6
LEGS see Livestock Emergency Guidelines and
        Standards
Live Export Accreditation Programme (Australia)
        197
Livestock Emergency Guidelines and Standards
        (LEGS) 139–40

livestock protection dogs (LPDs) 116–18
LPD *see* livestock protection dogs

malnutrition *see* thirst, hunger, malnutrition
methyllycaconitine (MLA) 85–6
milk sickness 82
monofluoroacetate (MFA) 85

neonatal mortality 6, 157–8
    animal welfare and 158–9
    birth difficulties/injuries 160–1
    in camels 178–9
    in cattle 168–72
    causes/risks in different systems 159–62
    in goats 177–8
    hunger and 159–60
    maternal behaviour and 159, 161, 162
    options to improve survival/enhance animal
        welfare 180–1
    in pigs 172–7
    predation and 161–2
    in red deer 179–80
    in sheep 162–8
    thermal stress and 160
nightshades (*Solanum*) 82–3
normal behavior 2
    grazing/management of extensive systems 67–8
    under thermal stress 19–20
North American Model for Wildlife Conservation
    109

OIE *see* World Organisation for Animal Health
One Health approach 149–50

pain, injury, disease 2
    control of 148–9
    during slaughter 198
    emerging infections 150
    examples of specific diseases 143–8, 151
    experience of 79–80
    grazing/management of extensive systems 67
    hazards affecting disease 137–41
    mitigating factors for disease risk 141–3
    neonatal 160–1
    spread of disease 148
    under thermal stress 18–19
    see also health
pigs
    heat stress 47
    neonatal mortality 159, 161, 172–7
plant secondary metabolites (PSMs) 78, 79, 88,
    90
PLF *see* precision livestock farming
poisonous plants 4–5
political conflict 139
precision livestock farming (PLF) 29–30, 151

predation 5–6
    current situation 105–8
    definition 103
    future prospects 125–6
    global trends 113–16
    historical experience 103–5
    lethal impacts of 108–13
    methods for dealing with 105, 106–7, 123–5,
        142
    neonatal 161–2
    non-lethal impacts 118–23
    reducing risk of 116–18
PSMs *see* plant secondary metabolites
pyrrolizidine alkaloids (PAs) 83–4

rangelands 1–2
    challenges 8
    degradation 4, 5
Rangelands Self Herding (or Rangelands Self
        Shepherding) 48
rayless goldenrod (*Isocoma pluriflora*) 82
red deer 179–80
reindeer 134, 151
Rift Valley fever 139
rinderpest (cattle plague) 139, 145–6

salmonellosis 139
saponins 84, 86, 88
sheep
    effect of four-wheel-drive motorbikes 151
    effect of predation on 112–14, 117–18
    foraging ability 59, 64
    grazing distribution 60
    neonatal mortality 162–8
    number of animals in US 112
    stocking rates 57
    thermal stress 11–12, 17, 21, 195
    transportation of 189, 191–2, 193, 194–6
    water balance 47
sheep scab 144
spillover host 148
stocking rate/density 4, 56–7, 142, 196
stockpersons 135–6, 140–1, 142–3, 153, 192
swainsonine plants 80–1, 87

temperature humidity index (THI) 15–16, 21
thermal stress 3, 11–12, 28–30
    animal responses to increased heat 43
    animal welfare under 16–20
    gene expression regulation under 22–4
    impact on animal productivity 20–2
    measuring 15–16
    neonatal mortality and 160
    strategies to reduce impact in ruminants
        24–8
    thermoregulation in ruminants 12–15

THI *see* temperature humidity index
thirst, hunger, malnutrition 2
    grazing/management of extensive systems
        62–6
    under thermal stress 17
ticks 61, 66, 136, 142, 143
toxic plants 78–80, 90, 138
    areas for future research 91–2
    birth defects/reproduction 86–7
    climate change/animal welfare 88–9
    effects on gastrointestinal tracts 82–3
    hepatic injury/photosensitization 83–4
    improving genetic fit of animals 87–8
    mitigation of effects 89–90
    mycotoxins 84–5
    myopathies 82
    neurological effects 80–2
    sudden death 85–6
    toxic minerals 87
transport systems 7, 188–9
    assembly depots 193
    by air 191
    by droving 189
    by sea 189–90, 193, 194–5, 196
    future research 197–8
    on land 190
    loading/unloading 193
    long-distance 191–3
    methods 189–91
    mortality/stress rates 193–6

quality of journey 192–3
standards 196–7
tsetse fly 144–5

water
    ambient temperature 40
    availability 3–4, 29–30, 37–8, 42–3, 138
    avoiding dehydration 43–6
    behavioural responses to temperature increases 43
    challenges in extensive systems 53–4
    conserving body water 43–6
    epigenetic responses 47–9
    essential nature of 49
    evaporative losses 41–2
    feed intake/forage type 40–1
    grazing/animal distribution 53–4
    influencing grazing behaviours 48–9
    management of livestock 46–7
    quality 42–3
    salt in 47–8
    use, conservation, management 39
    water balance 38
water hemlock 81–2
WCI *see* wind-chill index
white snakeroot (*Ageratina altissima*) 82
wind-chill index (WCI) 16
World Organisation for Animal Health (OIE) 145–6,
    197

zoonoses 132, 139, 146, 147, 149, 150, 153